WHITLEY STRIEBER'S HIDDEN AGENDAS

Whitley Strieber brings to light what others hope to conceal in WHITLEY STRIEBER'S HIDDEN AGENDAS. One of today's most respected names in the field of paranormal research, Strieber gathers together the best available evidence on such controversial issues as conspiracy theories, UFOs, close encounters, and other unexplained phenomena. Drawing on his vast knowledge and experience, spanning the globe and exploring a wide variety of subjects, Strieber brings us unprecedented access to information from reliable sources—revelations that will rock our deepest beliefs and may open a door to worlds other than our own.

ACROSS THE GLOBE,
THE EVIDENCE POINTS TO ONLY ONE CONCLUSION:
WE ARE NOT ALONE.

SOCORRO, NEW MEXICO, 1964
Police officer Lonnie Zamora sees a peculiar light in the sky, a roar, and a flash. Rushing to investigate what he believes to be a plane crash, he witnesses not only evidence of a UFO, but a sight even more disturbing.

CANARY ISLANDS, 1976
At least fourteen witnesses from the Spanish Navy report a luminous object moving offshore. And near the town of Las Rosas, several residents, including a physician, have a chilling encounter of the third kind.

BRAZIL, 1986
On May 19, the Brazilian Air Force confirms sending six planes to chase twenty UFOs seen and tracked by ground radar. Two days later eleven more UFOs return . . . and at that point all files are closed to the public.

Find out more in . . .

UFO BRIEFING DOCUMENT

UFO

BRIEFING DOCUMENT

The Best Available Evidence

BY DON BERLINER

WITH
MARIE GALBRAITH
AND
ANTONIO HUNEEUS

A DELL BOOK

Published by
Dell Publishing
a division of
Random House, Inc.
1540 Broadway
New York, New York 10036

Grateful acknowledgment is made to the following for permission to
reproduce Petit-Rechain UFO photo, copyright © 2000
Guy Mossay/Artist Rights Society, NY/SOFAM, Brussels.

Dell books may be purchased for business or promotional use or for
special sales. For information please write to:
Special Markets Department
Random House, Inc.
1540 Broadway
New York, NY 10036.

ISBN: 0-440-23638-X

Printed in the United States of America

Published simultaneously in Canada

First published in 1995 by the UFO Research Coalition

June 2000

10 9 8 7 6 5 4 3 2 1
OPM

CONTENTS

December 15, 1995

To whom it may concern:

We believe that this Briefing Document on Unidentified Flying Objects presents the best available evidence for the existence of UFOs. Although just a brief sample of the scientific and military evidence available worldwide is given, it represents some of the most *carefully documented* incidents.

While several governments of the world have dealt with this problem, as you can see in the enclosed report we think that these governments should make available now all the UFO evidence they have collected, for a thorough and open inquiry by the scientific community.

The political constraints that imposed the rule of secrecy during the Cold War are no longer justified and the solution to the UFO mystery may represent both a scientific and social breakthrough.

We, the undersigned, endorse the information contained in this Briefing Document as the best available evidence from open sources.

CUFOS (Center for UFO Studies):
  President
  Dr. Mark Rodeghier

FUFOR (Fund for UFO Research):
  Chairman
  Mr. Richard H. Hall

MUFON (Mututal UFO Network):
  International Director
  and President
  Mr. Walter H. Andrus

ACKNOWLEDGMENTS

Without the enthusiastic assistance of many people, the creation of this Briefing Document would have been far more difficult, if not impossible. While there are too many for us to thank individually, some deserve special recognition:

Laurance S. Rockefeller, for his vision and support, financial and otherwise, and George Lamb, for his day-to-day interest and for serving so effectively as liaison for Mr. Rockefeller.

Marie "Bootsie" Galbraith, for the original idea and for hundreds of hours of turning it into reality. Sandra Wright, for making her BSW Foundation available as the umbrella under which all the work could be done. Tina Nighman, for applying her talents and good humor to a wide range of administrative assistance.

The leadership of the UFO Research Coalition: the Center for UFO Studies, the Fund for UFO Research, and the Mutual UFO Network, for their cooperative efforts and total support.

Major General Wilfred De Brouwer, Deputy Chief of the Royal Belgian Air Force; Dr. Claude Poher, founder of the Groupe d'Etudes des Phénomènes Aerospatiaux Non-identifiés (GEPAN); Jean-Jacques Velasco, Director of the Service d'Expertise des Phénomènes de Rentrées Atmosphériques (SEPRA); the Société Belge d'Etude des Phénomènes Spatiaux (SOBEPS), and internationally recognized UFO authorities Stanton T. Friedman and Timothy Good, for generously giving their time and help.

INTRODUCTION

Whitley Strieber

1944: a clear afternoon over northern Italy, hard light, empty sky. American bombers are coming in high over the Po valley on a mission to attack German defensive positions. Below, the Luftwaffe scrambles a few ramshackle Me-109s. As the German fighters ascend, the pilots notice glowing balls of light pacing them just off their wingtips. Meanwhile, the American flyers above see the same thing—balls of light darting among their formations.

Pilots from both sides watch the objects warily. All assume that they are some sort of enemy secret weapon or tracking device.

It is not until after the war that it is realized the "foo-fighters," as the Americans called them, were not created by either side. Early investigation of the objects led nowhere, and the matter was forgotten.

Then came June 24, 1947. Kenneth Arnold was flying his single-engine airplane near Mount Rainier in Washington State, when he was startled by a flash. He then observed nine silvery craft slipping through the air, approaching Mount Rainier at a rapid rate of speed.

His report caused a sensation in the nervous postwar United States. It was the beginning of a sighting wave so massive that it was attributed by an uneasy press to hysteria. However, the Air Force was not so quick to dismiss the sightings. Indeed, at high levels there was considerable concern. Air Force intelligence was informing high command that the observations were "not at all imaginary." And that "something is really flying around."

Answers were needed, and a project was devised to get them. Thus, in 1948 began Project Sign, and with it the start of the UFO controversy. Sign was inconclusive. It was followed by Project Grudge. In 1955 the Air Force released a

study of 3,200 UFO reports, saying that almost half were unexplained. It continued with its study, this time under the
name Project Blue Book. In 1967 it contracted with the University of Colorado to produce a study of the thousands of
cases it had now accumulated.

When this study was released, its conclusion was that the
UFO mystery was not worth scientific investigation. This
was a startling and unexpected outcome, especially in view
of the fact that thirty percent of cases had continued to defy
explanation.

Members of the Blue Book Committee resigned in pro-
test. There were books written to refute the conclusions in
the report. But its recommendations became official policy.
The Air Force ceased to collect UFO data. Scientific institutions declined to address the matter.

Nevertheless, UFO waves continued. Through the seven-
ties and the eighties, sightings became more numerous and
detailed. People began to claim close encounters with apparent aliens. But nobody knew the truth because there was no
systematic study taking place. The question of whether UFOs
were even real, let alone spacecraft from other planets, was
officially ignored as a matter of national policy.

Not everybody was convinced that this was an appropriate posture, and there was continuous pressure for the government to get back into the UFO-reporting business and for
science to take another look at the situation.

It reached a point, by the mid-nineties, when the issue
was effectively forgotten by science.

The community of professional UFO researchers was
concerned not only that this crucial issue was being ignored
but also that a valuable resource of new knowledge was
being debased and confused by wild public speculation. A
decision was made to attempt to preserve the record and
reintroduce the issue into the court of world opinion.

It was determined that a good approach might be to create a compendium of the best cases in history and send them
to everyone who might make a difference. Publication was

to be restricted to a small group of highly placed officials and scientists only.

Thus was born the UFO Briefing Document. It quickly became a matter of legend. What cases did it cover? Why was it thought to be so convincing? What secrets did it reveal?

It was funded by Laurance Rockefeller and prepared under the supervision of his associate Marie Galbraith. Long-time UFO investigators Don Berliner and Antonio Huneeus, acknowledged to be two leading experts in the field, assembled the cases. The BSW Foundation produced and distributed the document.

It was sent to prominent politicians, world leaders, and scientists in 1996. Rockefeller and his group waited expectantly for a response, but it was virtually ignored by the community of leaders who received it.

The document offers powerful evidence that UFOs are real. Indeed, the evidence presented here is overwhelming. Every skeptical objection is met: There are ample professionally trained witnesses involved; there are multiple witnesses; there is photographic support in some of the cases.

The document does not ask the reader to "believe in" UFOs. What it asks for is official recognition of the fact that the phenomenon is not understood and an appropriate scientific effort to explain it.

Some of the cases are truly startling, and one comes away from reading them with a deep sense of concern over why such an obvious mystery draws only official silence. It turns out that many highly skilled professional observers have seen UFOs and that there is almost irrefutable physical evidence associated with some UFO events.

A UFO chase is recorded on radar. Naval personnel photograph an apparent alien spacecraft. UFOs cause air-base alerts. An astonishing sighting wave stuns an entire country.

So why isn't this phenomenon—undeniably real and undeniably mysterious—treated like other unknowns? It isn't

that the techniques of analysis are unavailable. There are
dozens of different ways to detect and analyze UFO activity,
most of which have never been tried. For example, either
satellite or ground-based cameras or both could do continuous sky searches that would detect and record essentially
everything in motion. Cameras capable of this already exist,
and a ground-based system that would cover the entire
United States could be deployed for under two million dol-
lars. Satellites that are easily capable of photographing objects
in space are already in place. All that would be necessary is
to make relatively minor changes in their orientation and ad-
justments to accomplish complete monitoring of near-earth
space. In addition, the North American Air Defense Command possesses an electro-optical monitoring system that is
capable of detecting an object the size of a Ping-Pong ball
at a distance of fifty thousand miles in space, even a fast-
moving one.

So if all the equipment is so readily available, why is
nothing being done?

The reason is that even well-informed scientists dis-
believe the UFO information so completely that they do
not consider it worth pursuing. Their ignorance is supported
by a government that remains as secretive as it was during
the Cold War. Virtually all the major Cold War acts—the
National Security Act of 1947, the National Space Act of
1958, and a host of other secrecy acts and executive orders—
remain the law of the land. Although a declassification process exists, there is evidence that an effort has been made to
actually hide UFO information, to move it outside the system altogether in order to shield it from declassification.

The UFO Briefing Document presents damning evidence
of government secrecy, narrowing the whole matter down to
a single, extremely telling court case that makes it all but impossible for a thinking person to deny that the government
keeps some UFO documents very, very secret.

Because of government secrecy, scientific indifference,
and fifty years of press debunking, the whole UFO question

has been marginalized. The Unidentified Flying Objects Briefing Document seeks to reopen it, and the proof that it assembles is overwhelmingly convincing. UFOs are real. To all appearances they are under intelligent control. And they come close to us at times, as some of the most shocking photographs in this document reveal.

The important question is not whether or not UFOs exist but why our society has responded in such a strange and inappropriate manner to their presence. The United States Air Force has not acted in a reasonable manner, especially in view of the fact that adequate equipment existed during the term of Project Blue Book to answer definitively the question of whether or not the objects were real. However, Blue Book was exclusively directed toward sighting reports. No effort was made by the Air Force, nor has such an effort ever been made, at least not publicly, to go out into the field and gather evidence in an organized and scientifically sound manner.

This should certainly have been done prior to dropping the question. The fact that the phenomenon is elusive and hard to photograph is not a realistic obstacle for determined, well-equipped investigators. High-speed cameras that could cover large areas of sky were developed in the forties and were readily available in the fifties. But no Air Force project ever made use of them for the purpose of attempting to track UFOs then, and nothing is being done now.

Because we have failed to apply proper scientific resources to this issue, we remain in a state of ignorance about UFOs. All we *do* know is that they are enough of a physical phenomenon to leave not only eyewitness reports but a substantial body of photographic and video evidence and reports of electromagnetic effects such as those that appear in the Briefing Document.

Since they give every evidence of being under intelligent control, and quite possibly being alien spacecraft, it would seem that an investigation is imperative. But it does not happen.

Why not?

Superficially the answer would appear to be that fifty years of official denial has caused a cultural bias to develop against the phenomenon. The refusal of the government to admit that UFOs are a genuine unknown has resulted in a parallel refusal by funding organizations to pay for the sorts of scientific studies that would be required to solve the mystery and has caused the National Academy of Sciences to actually prohibit such studies in institutions under its authority.

Because of the official denials and the lack of scientific study, there has developed a bias against the subject within the intellectual community and the media. UFO-related stories, no matter how extraordinary, do not generally achieve coverage beyond the level of local newscasts, if at all. The bias against UFO reporting is stronger in the print than in broadcast media, to the point that it is highly unusual when anything related to UFOs gets coverage in newspapers and magazines, except to be rejected and debunked.

When the UFO Briefing Document was given to world leaders, the press was informed. The results were so negative that some of the parties responsible for the document are now unwilling even to be interviewed about it.

On April 8, 1996, the New York *Observer* published a report on the release under the headline "Rockefeller Greets Aliens! A Rich Guy's UFO Dream." The tone of the article is somewhat less negative than the headline would suggest, describing the report as a "dramatic document."

After the distribution was completed, it seemed as if the report was going to be filed away by the community of leaders who had received it: 1997 and 1998 passed with little response. However, in 1999 an event occurred that began to make it seem as if the UFO Briefing Document had been taken as seriously as its creators had hoped, at least in some countries.

Following the lead of the United States, most countries ignore UFO reports. A few have a different approach, among them Belgium, Ecuador, Chile, and France. Ecuador has followed the French model, which has been to create and maintain an official center for the collection of UFO reports. Despite this somewhat more open approach, though, there is no organized and ongoing sky search, no organized system for gathering possible material residue and artifacts, no effort to detect signals. However, the Briefing Document has not been entirely ignored.

In July 1999 a document entitled "Les Ovnis et la Defense: A quoi doit-on se prepare?" (UFOs and Defense: For What Should We Prepare?) was published in France. The ninety-page report was the result of a study conducted by an independent group of former "auditors" at the Institute of Higher Studies for National Defense and by qualified experts from numerous different scientific disciplines, as well as the military. The institute is the French equivalent of the Rand Corporation.

Prior to its publication, the report was presented to the French president Jacques Chirac and prime minister Lionel Jospin.

This report is the closest any government has ever come to officially acknowledging that UFOs represent a real mystery and a serious issue. It acknowledges its debt to the Briefing Document and thus not only confirms the document's importance but also vindicates its sponsors and authors. It is to be hoped that French diplomatic efforts will follow the release of the report and that the United States will respond in a reasonable manner. Certainly it cannot be claimed that United States officials are ignorant of the truth about UFOs. This Briefing Document has been sent to the White House, to every member of Congress, and to numerous other officials who might have reason to concern themselves with it.

The French report was written by the most distinguished group of officials and scientists ever to write a UFO report. Prefaced by General Bernard Norlain of the French Air Force, the former director of the Institute for the Higher

Studies of National Defense, it begins with a preamble by André Lebeau, the former president of the National Center for Space Studies, the French equivalent of NASA.

The group that collectively authored the report calls itself the Committee for In-Depth Studies, or COMETA. It is presided over by General Denis Letty of the Air Force. Included among its members are General Bruno Lemoine of the Air Force; Admiral Marc Merlo; Michel Algrin, Doctor of Political Science; General Pierre Bescond, armaments engineer; Denis Blancher, the chief national police superintendent of the Ministry of the Interior; Christian Marchal, the chief engineer of the National Corps of Mines and Research Director of the National Office of Aeronautical Research; and General Alian Orszag, Ph.D., physicist and engineer for armaments.

Had such a blue-ribbon committee been assembled in the United States, there may have been a small amount of press coverage in this country. As matters stand, there was no coverage of this report in the English-speaking press at all, not in the United States, the United Kingdom, Canada, or Australia. Indeed, the failure on the part of the press to cover this story speaks volumes about just how effective the United States government's policy of official denial has been over the years.

The report makes frank reference to the veil of secrecy that has been dropped over the UFO phenomenon by the United States government, calling it "strange." It stresses the important role that private organizations in the United States have played in preserving and publishing UFO information and particularly mentions this Briefing Document as a critical source of knowledge.

Like the Briefing Document, and like all other serious UFO studies, the French report addresses the questions of what UFOs are, how they function, and what might be the motives of their occupants. It is at this point that defense issues are raised. It is noted that numerous instances exist of

UFOs flying over nuclear missile sites, and reference is made to a quite amazing multiple-witness case at a Russian missile base, which is also covered in the Briefing Document. The United States' attitude of denial and secrecy is described as "most strange," and it is noted that the level of secrecy surrounding the subject in the United States is getting higher.

Given the powerful cases reported herein, the response of the French and now also the publication of Dr. Peter Sturrock's book *The UFO Enigma* could be a sign that change is finally taking place. Dr. Sturrock, professor emeritus of physics at Stanford University, has based his book on a report that was distributed to the press in 1998 and was published in 1999. The report, entitled "Physical Evidence Related to UFO Reports," was originally published as the proceedings of a workshop held with the support of Laurance Rockefeller between September 29 and October 4, 1997. It was developed by Dr. Sturrock and a committee of distinguished scientists.

On June 29, 1998, a few days after the Sturrock document was released, there was a change in the approach that at least one major newspaper has been taking to the UFO question. The *Washington Post* ran a serious and straightforward story about it. The *Post* commented that "the panel suggests that the scientific community has suffered a failure of curiosity regarding UFOs." It goes on to report that "a sampling of scientists outside the panel expressed surprise that a topic with such a high 'giggle factor' might be reincarnated for serious study, possibly further blurring the lines between legitimate research and the 'lunatic fringe.' Some said they would never comment on the touchy topic. . . ."

Thus the scientific community as a whole appeared to be responding more to the cultural bias against the UFO mystery than to the increasingly strong evidence that it is worthy of study.

The Briefing Document was seminal to the production of the Sturrock report and the book that followed. Some of the cases referenced in the book are, once again, from the Brief-ing Document. The French COMETA report takes the posi-tion that UFOs are probably of extraterrestrial origin, as does the Briefing Document, but the Sturrock report is more care-ful, concluding only that the evidence requires further study.

In the end the question must be asked: Given the power of the evidence presented in the Briefing Document and elsewhere, why is there all the denial and secrecy? What sane reason could the government have for a cover-up? And why do scientists perceive only a "giggle factor" when what is arguably the most important issue in human history is raised?

It is unquestionably true that there would be no "giggle factor" if the government had responded from the beginning in an open and straightforward manner. But it did not. In-stead, its continued and vociferous denials spread disbelief to the point that the scientific community, with its denials, is at odds with the popular will. The credibility of science has been eroded along with that of the government.

Because science has failed to respond, people have turned to other authorities for explanations. The result is that soci-ety has been flooded with half-baked theories and superstitions about UFOs having little or no grounding in fact.

Thus, UFO secrecy has not only denied science the chance to make what is probably the discovery of the ages, it has corroded the intellectual content of the culture in general and opened the popular imagination to a level of superstition that gives every evidence of forming itself, in time, into a new cult or religion.

Why would it be like this? What is the origin of all this secrecy? Perhaps the government realizes that it has been extremely irresponsible and wishes to conceal its policy fail-ure as long as it can. Or maybe it did something that has been profoundly detrimental to contact, and this must be con-cealed. Maybe, also, the aliens themselves enforce the se-crecy, and therefore the whole peculiar edifice seems so directionless and arbitrary because there is no human policy behind it. Maybe the government is not covering up the UFO enigma intentionally but is being compelled to do so by an outside force—a possibility that may not be as unlikely as it seems.

On May 6, 1977, Drs. T.B.H. Kuiper and M. Morris published a paper in *Science* entitled "Searching for Extraterrestrial Civilizations." The paper bore the subhead "The search for extraterrestrial intelligence should begin by assuming that the galaxy has been colonized." They proceed to point out that if any intelligent species can travel through interstellar space, it is almost inevitable that the whole galaxy is already colonized. This would explain Professor Hermann Oberth's 1954 comment "There is no doubt in my mind that these objects are interplanetary craft of some sort. I am confident that they do not originate in our solar system, but they may use Mars or some other body for a way station." In an interview by Briefing Document author Antonio Huneeus, Professor Oberth, who is regarded as the father of German rocketry, stood by his statements.

Perhaps he was aware that a study conducted by another pioneer, Dr. John Von Neumann, regarded as the father of cybernetics, suggested that a spacefaring species could colonize an entire galaxy the size of our own in a million years even if they could not travel in excess of the speed of light.

Thus there are credible assertions on the part of prominent scientists that support the idea that aliens may already be here. But this still does not get to the issue of secrecy . . . or does it?

In their paper, Kuiper and Morris postulated that aliens, upon discovering us, would almost certainly maintain a rigorous wall of secrecy around themselves. They state: "In the postindustrial society, knowledge is the most valued resource. . . . Thus we suggest that knowledge, in a general sense that encompasses science and culture, is likely to be most highly prized by an advanced civilization. . . . Before a certain threshold is reached, complete contact with a superior civilization (in which their store of knowledge is made available to us) would abort further development through a 'culture shock' effect. If we were contacted before we reached this threshold, instead of enriching the galactic store of knowledge, we would merely absorb it."

They conclude that "by intervening in our natural progress now, members of an extraterrestrial society could easily extinguish the only resource on this planet that could be of any value to them," which is the free and independent growth of human knowledge.

If aliens are here, then, it may be that they themselves are the architects of the secrecy. This could be the only explanation available for continued governmental resistance to telling the truth, even in the face of documents like this, which offer completely irrefutable evidence that UFOs are real.

PART 1
OVERVIEW

GOVERNMENT SECRECY

In a democracy, the decision of where to draw the line between a citizen's right to know and the government's right to secrecy for national-security reasons must be made by appropriate members of the society. This issue has become the focus of much attention today and is especially relevant to an ongoing discussion, both inside and outside Congress, regarding UFO phenomena.

For obvious reasons, military services and the intelligence agencies must maintain a certain amount of secrecy. However, in recent decades, and especially since the end of the Cold War, many observers believe that the use of government secrecy has become excessive.

The power of government employees to restrict access to reports that they write by classifying them "confidential," "secret," or even "top secret" is often absolute. Once these reports are classified, they can be declassified only by the originator or by a special procedure that moves along at a glacial pace. Nor does the Freedom of Information Act (FOIA) help very much. It does not apply to most classified material. Meanwhile, our criminal statutes protect against the unauthorized revelation of classified materials.

Secrecy, like power, lends itself to abuse. Behind the shield of secrecy, it is possible for an agency or service to avoid scrutiny and essentially to operate outside the law. Accountability to the taxpayers and to the Congress can be conveniently avoided.

The vast majority of people employed by the U.S. government do not have access to classified information. Even those with secret and top-secret clearances will not have access to *all* highly classified information. Furthermore, it is doubtful whether *any* member of Congress can have access to *all* such information. Given the size of the government bureaucracy and high degree of compartmentalization that exists within it, it is conceivable that even the President himself is not *fully* briefed on matters classified as "above top secret." Such information, allowing access only on the strictest "need to know" basis, is not necessarily given to senior elected officials who come and go and can therefore be regarded as temporary, political, and unreliable.

Such is the case for top-secret UFO information. In 1980, for example, researchers requesting information through the FOIA learned of the existence of 156 top secret UFO-related documents held by the National Security Agency (NSA). This lead was not found through the NSA itself but through internal references in UFO-related documents held by *other* government agencies. When the researchers filed a FOIA request for the 156 NSA UFO documents, they were denied access to all of them. They appealed, but Judge Gerhard Gesell of the First Federal Court, District of Columbia, after reviewing the 21-page written argument submitted by the NSA, denied their appeal. The 21-page summary was later released, but even in this "summary" most of the information was blacked out.[1] Such action seems inconsistent with a government that *officially* downplays the existence of true UFOs, and *officially* states that there is no threat to national security.

In the case of UFO phenomena, the question must be asked: What would give an unelected government official the right to keep this information to himself, thereby depriving the rest of the world of possible knowledge of almost inconceivable magnitude and consequence? Such elitism by the officials of any government, much less a government based on the principles of democracy and individual rights, is a gross injustice not only to its own people but to all people.

At issue in this case is access to knowledge perhaps so profound that it affects not only our very perspective on

_____
[1] Judge Gesell Ruling re National Security Agency, November 14, 1980.

man's place in the universe, but also perhaps his continued presence on this planet. If the UFO phenomena are real, we have clear evidence that an unknown technology is at work, whose potential could be enormous for the good of mankind— a potential source, for example, for useful energy benign to the environment.

To *acknowledge* the enormous gap between our present understanding of science and what is now being evidenced would provide the urgently needed challenge to the scientific establishment to examine where some of its basic assumptions might be faulty and to move beyond them.

Is it possible that a few privileged individuals have access to this information while denying it to the electorate for "national security" reasons, so that it can be privately studied? In a democracy, should not this decision be made by our *elected* officials and be based on informed discussion?

> *UFO research is leading us kicking and screaming into the science of the 21st century.*
>
> *I have begun to feel that there is a tendency in 20th Century science to forget that there will be a 21st Century science, and indeed a 30th Century science, from which vantage points our knowledge of the universe may appear quite different than it does to us. We suffer, perhaps, from temporal provincialism, a form of arrogance that has always irritated posterity.*
> [From a letter by Dr. J. Allen Hynek to *Science* magazine, August 1, 1966.]

Dr. J. Allen Hynek, Northwestern University astronomer; scientific consultant on UFOs to the U.S. Air Force from 1948 until 1969. Founder of the private Center for UFO Studies in 1973.

THE CASE FOR UFO REALITY

As long as men and women have talked about strange sights in the skies, two primary questions have been asked about what have come to be called Unidentified Flying Objects:

1. Are they real, or are they just honest mistakes?
2. If they are real, could they be ships from some other world?

In the twentieth century, it started with the "foo fighters" of World War II: glowing balls that flew in formation or "played tag" with military airplanes over Europe and the Pacific. Suspected of being prototype enemy weapons, they never displayed hostility, and when the war was over they were all but forgotten.

In 1946, the Scandinavian countries reported many hundreds of "ghost rockets" that flew low and silently, and often slowly. Efforts to blame them on nearby Soviet tests of captured German missiles failed when it was learned that no such tests had taken place.

In America, the first major wave of sightings of "flying discs" began in the early summer of 1947. Within two weeks, at least 1,000 sightings were recorded of fast, silvery discs seen in the daytime. The first military studies concluded that they were real and of unknown nature and origin.[2]

From then on, UFOs seemed to fly at will over all parts of the world: fast and exotic, untouchable and unproven. By the 1990s, there had been over 100,000 reported sightings, many by airline pilots, military pilots, and other qualified witnesses.

---

[2] Memo from Lt. Gen. Nathan Twining, Commanding General of the Air Materiel Command, Wright Field, to Gen. Spaatz, Commanding General of the U.S. Army Air Forces, September 23, 1947.

Despite the steady accumulation of a vast quantity of information about the appearance and behavior of UFOs, little light has been shed on the two questions posed at the beginning. The armed services and universities, as well as private groups and individuals, have devoted a great amount of time to investigating UFOs, yet there is no consensus about their nature, origins, or purpose.

Still, if a close look is taken at the best available evidence, it is possible to deal with what is known about UFOs, and what may reasonably be assumed. The point we will make is that the evidence to support the conclusion that UFOs are unknown aircraft/spacecraft seems to be overwhelming.

Visual Evidence

Most of what is "known" about UFOs comes from individuals' descriptions of what they say they saw. If the individuals are reliable and knowledgeable about the sky, the information stands a good chance of being useful. This is the source of the case's "credibility," one of the two primary criteria recognized by the late Dr. J. Allen Hynek, long a consultant on UFOs to the U.S. Air Force, and later the founder of the private Center for UFO Studies.

Dr. Hynek's other criterion is "strangeness," meaning the extent to which a reported observation differs from normal airplanes, satellites, meteors, etc. A large aluminum-looking sphere that maneuvers violently and changes speed abruptly rates higher for "strangeness" than a somewhat peculiar light seen in the night sky.

It is the reports that rate highest in both "credibility" and "strangeness" that form the heart of the UFO mystery. Are they indeed convincing observations of unknown aircraft/spacecraft, or are they merely strangely shaped clouds or balloons seen under unusual lighting conditions, or some other natural or man-made phenomena?

Radar Evidence

Radar has played a major role in UFO sightings, repeatedly confirming the presence of something unidentified that responds to radar much as an airplane does. Clouds and other weather phenomena show up on radar, but any experienced operator can tell the difference between weather and something solid.

One popular explanation for radar/visual reports is temperature inversion. This was first brought to public attention following two nights of UFO sightings over Washington, D.C., in 1952. Inversions, which cause mirages, probably never caused these or any other UFO reports. According to a 1969 study by the Air Force Environmental Technical Applications Center, the conditions needed to produce the UFO-like effects attributed to inversions cannot exist in the Earth's atmosphere.[3]

One case of thoroughly investigated radar/visual UFO sightings occurred in Belgium and Russia in 1995. Military jet interceptors were launched following observations from the ground. Ground-based and airborne radars then confirmed what was being seen visually, including high speeds and violent maneuvers far beyond the capability of the best modern warplanes. In both countries, high government officials admitted they were baffled.

While the human eye can be fooled, and radar can be fooled, it is considered extremely unlikely that both can be fooled, in exactly the same way, at exactly the same time. Thus, radar/visual reports rate among the most convincing of all types of UFO sightings.

---

[3] Menkello, F.V., "Quantitative Aspects of Mirages," USAF Environmental Technical Applications Center, 1969.

Physical Evidence

UFOs have been seen high in the sky, near to the ground, on the ground, and even rising from water. If some UFOs have landed, it is reasonable to suspect that some of them may have left traces behind, and indeed that is the case. Imprints, residues, charred and broken tree branches and rocks are among the bits of evidence claimed for UFO landings. Furthermore, under microscopic examination, some residues exhibit strange and unusual characteristics.

Perhaps the most well-known example of a physical trace case in the United States occurred in 1964 near Socorro, New Mexico, where a policeman reported seeing an egg-shaped craft sitting on slender legs in an open field. When it had flown away, he and a second policeman inspected the area where it had been parked and found depressions in the dirt, as well as still-smoldering, blackened shrubs. The sighting was investigated within two hours by men from U.S. Army Intelligence and the FBI, followed a day later by the chief civilian scientific consultant to Project Blue Book (the official Air Force investigation of UFO sightings). All agreed that the primary witness was highly reliable. Later, the final director of Blue Book called this case the most puzzling of the approximately 12,500 in his files.[4]

One of the best documented examples of a physical trace case in Europe occurred in Trans-en-Provence, France, where a farmer reported seeing a saucer-shaped craft land on his property and then fly away after a short while. Physical traces left on the ground were collected by the police within twenty-four hours and later analyzed in several French government laboratories. Microscopic analyses revealed anomalous biochemical and electromagnetic effects on the soil and vegetation. The director of the Service d'Expertise des Phénomènes de Rentrées Atmosphériques (SEPRA, formerly called GEPAN)

---

[4] Steiger, Brad, ed., *Project Blue Book,* Ballantine Books, 1976.

at the National Center for Space Studies (CNES) describes this case as one of the most puzzling UFO cases in the French government files.[5]

Government Statements

The involvement of the American government in the UFO mystery has long offered its own set of questions. Known investigations have produced ambiguous results, and explanations offered for specific cases have frequently been at odds with scientific reasoning. Sometimes, little-publicized official statements have supported the position that UFOs are real and unexplained.

Sometimes statements not intended for the public have been brought to the surface by UFO researchers:

> *July 30, 1947: This "flying saucer" situation is not all imaginary or seeing too much in some natural phenomena. Something is really flying around.[6]*

> *Sept. 23, 1947: The phenomenon reported is something real and not visionary or fictitious.[7]*

> *Oct. 28, 1947: It is the considered opinion of some elements that the object [sic] may in fact represent an interplanetary craft of some kind.[8]*

---

[5] GEPAN, *Note Technique No. 16, Enquête 81/01, Analyse d'une Trace,* Toulouse, March 1, 1983. (English translation published in the *MUFON UFO Journal,* March 1984.)

[6] Air Force Base Intelligence Report, "Flying Discs," AFBIR-CO, July 30, 1947.

[7] Twining, ibid.

[8] Draft Intelligence Collections Memorandum issued by Brig. Gen. George Shulgen, Chief of the Air Intelligence Requirements Divi-

Dec. 10, 1948: It must be accepted that some type of flying objects have been observed, although their identification and origin are not discernible.[9]

In 1948, the U.S. Air Force opened a publicly known UFO investigation called Project Sign. Later, it became Project Grudge and finally Project Blue Book. In 1955, the U.S. Air Force released a study of 3,200 UFO reports it had received between 1947 and 1952. The private Battelle Memorial Institute used the Air Force data to arrive at its own conclusions: of the cases for which there was some conclusion, almost 50 percent were either unexplained or doubtfully explained. Moreover, it was determined that the higher the qualifications of the witnesses, the harder it was to explain the reports in terms of common phenomena.[10]

In 1967, as Project Blue Book was coming under increasing attack from the press and the public, the Air Force contracted with the University of Colorado to make a final study of UFOs. In contrast to the totally negative statements of the study director, Dr. Edward U. Condon, the body of the final report showed that about 30 percent of the cases studied were left without explanation.

Comments on individual cases by University of Colorado scientists included:

This is the most puzzling case in the radar/visual files. The apparently rational, intelligent behavior of the UFO suggests a mechanical device of unknown origin as the most probable explanation.

---

sion of the Office of the Assistant Chief of Staff of the U.S. Air Force, October 28, 1947.

[9] U.S. Air Intelligence Report #100-203-79, "Analysis of Flying Objects in the U.S.," December 10, 1948.

[10] Air Force Project Blue Book, "Special Report No. 14 (Analysis of Reports of Unidentified Aerial Objects)," May 5, 1955.

All factors investigated—geometric, psychological, and physical—appear to be consistent with the assertion that an extraordinary flying object, silvery, metallic, disc-shaped, tens of meters in diameter, and evidently artificial, flew within sight of two witnesses.[11]

Following the recommendation of the University of Colorado, Project Blue Book was ended in late 1969, after almost twenty-two years of Air Force official investigations. It left behind approximately 12,500 case files, of which 585 were officially declared "Unknown." This means that the project staff felt it had sufficient *information* about a case but was unable to supply a full *explanation* of it.

Cases lacking sufficient information for meaningful analysis were kept separate. Furthermore, an official memo was released years later, under the Freedom of Information Act, that made it clear that "reports of unidentified flying objects *which could affect national security* . . . are not part of the Blue Book system." [Emphasis added.] Such reports "would continue to be handled through the standard Air Force procedures designed for this purpose."[12]

In summary, it is apparent that the evidence—visual, radar, and physical—strongly suggests that more than mistaken observations of conventional phenomena are involved in many UFO sightings. Witness testimony, backed up by official U.S. government documents, points toward the presence in the Earth's atmosphere of apparently manufactured craft that cannot be explained as mistaken observations of acknowledged aircraft, spacecraft, or atmospheric or astronomical phenomena.

---

[11] Gillmor, Daniel S., ed., *Scientific Study of Unidentified Flying Objects,* New York Times Books, 1969.

[12] Bolender, Brig. Gen. C.H., USAF, Memo re Project Blue Book, October 20, 1969.

The Case for Extraterrestrial UFOs

If UFOs are not anything known, then they must be unknown. What says "unknown" more powerfully than "extraterrestrial"? In the absence of any specific knowledge of even a single extraterrestrial civilization, there are no constraints on theorizing about the nature, technology, and behavior of one or more hypothesized alien cultures.

But are UFOs extraterrestrial? Lacking proof, we must deal very carefully with any answers. It remains a possibility that some or all of the otherwise unexplained UFO reports will someday be explained in terms of as-yet-unknown natural phenomena or secret, highly advanced man-made aircraft and/or spacecraft.

Nevertheless, there are impressive reasons for speculating about the extraterrestrial origin of some UFOs, namely their shapes and their performance.

Shapes of UFOs

Most UFOs observed in daylight, when shapes and details can best be seen, have been described as having simple geometric shapes: discs, spheres, cylinders, and, more recently, triangles.

Disc-shaped airplanes have been flown, but none is known to have exceeded 150 mph or to have other capabilities displayed by UFOs. Difficulties in stability and control have so far prevented any disc-shaped aircraft from getting beyond the stage of low-performance prototypes.

Spherical aircraft have so far been limited to gas-filled balloons, whose performance is at the bottom of the speed-and-maneuverability scales. Balloons can fly only as the wind blows and can be overtaken quickly by airplanes.

Cylindrical aircraft are unknown, as the lack of wings poses huge problems when it comes to such functions as taking off and flying level. Rockets and missiles are cylindrical and certainly are able to fly, but only as the result of great power in relation to their size. They can fly up only at launch, and on a ballistic curve on their way to a target.

Triangle is the shape of delta-winged airplanes, though the flight characteristics of triangular UFOs remove them from this category.

It is entirely possible that some radical military aircraft having one or more of these shapes are flying from super-secret test facilities. But this would have to be a recent development unable to explain sightings of such craft during most of the past fifty years.

Performance of UFOs

Even more striking than the shapes of UFOs is their performance: speed, acceleration, maneuverability, silence.

Speed. UFOs have been tracked on military radar traveling silently at several thousand miles per hour *well within the Earth's atmosphere.* An airplane attempting this would create an inescapable sonic boom before melting from friction with the air.

Extreme acceleration. Airplanes do not visibly accelerate in the air, though they show generally impressive acceleration during takeoff. Drag-racing cars and motorcycles accelerate in a manner obvious to even the least experienced observer. In the case of UFOs, airline and military pilots have reported that they fly at the same speed as an airplane, then display acceleration common only to antimissile missiles. Veteran pilots describe their observations with words like *astounding* and *unbelievable*.

Extreme maneuverability. While airplanes can perform abrupt maneuvers, these are generally seen only in air shows. Even then, such flying is more often described by the outside observer as *graceful* rather than *violent*, though the pilot may use the latter term. Impossibilities for airplanes (but not, apparently, for UFOs) include right-angle turns at high speed and zigzag flight.

Silent hovering. While helicopters and VTOL (Vertical Take-Off and Landing) airplanes can hover, they produce noises whose quality and volume positively identify them. UFOs, on the other hand, appear able to hover with little or no motion for long periods without any sound. This remains well beyond the state of known science, let alone technology.

Summary

The U.S. government, and many other governments, claims that although not all UFO reports can be explained, there is no evidence that Earth has been visited by aliens. Most scientists and leading journalists agree with this position. However, these same scientists believe that there must be many advanced civilizations on planets orbiting the billions of stars they estimate to exist in the universe. The gap between these two positions is generally explained by the *assumed* inability of even the most advanced society to travel the enormous distances separating Earth from even the nearest stars.

Yet, there are thousands of sightings of novel, high-performance craft in our skies, reported by highly skilled and experienced observers. There are also hundreds of other reports of craft seen on the ground, and sometimes of humanoid beings in their vicinity.

The great conflict between official positions and trustworthy observations constitutes the mystery of Unidentified Flying Objects. A possible solution to this mystery is the suggestion that the official position is based on an elaborate cover-up. If it is a cover-up, what then is being protected, and by whom?

The answers to these questions generally focus on the issue of national security and fear of the public reaction to an official disclosure of UFO reality and its extraterrestrial origin. The question of extraterrestrial intention and the frightening aspects of the alleged abduction phenomena could be extremely disturbing. However, many researchers believe that it is the science and technology behind the national-security veil that lies at the heart of the secrecy, and that:

Fallen discs are being reverse engineered, repaired and/or copied, and being tested.

The technology is so advanced that we can barely imagine the science behind it (which could be based on a fundamentally different understanding of gravity and electromagnetic fields).

Whichever nation masters this extraordinary technology will certainly be the most powerful nation on Earth.

In the opinion of those in control, the guarding of this technology for defense purposes far outweighs its potential value for other purposes—i.e., a nonpolluting, cost-efficient solution to our present energy and environmental crises.

THE UFO COVER-UP

There are two major elements to the UFO mystery: the UFOs themselves and the intensive efforts by the governments of the world to withhold information about them. Neither the nature nor the purpose of the governments' actions is clearly understood. But this policy dates back to the latter part of World War II, when UFO-like "foo fighters" were being reported by combat pilots.

A report about "foo fighters" is said to have been prepared in 1945 by the United States Eighth Air Force, but no copy has been seen by the public, despite the passing of more than half a century.[13] A year later, when "ghost rockets" were seen over Scandinavia, the Swedish government invoked secrecy, and only began to release information forty years later. When "flying saucers" appeared over the United States in the summer of 1947, only the most general information was made public, while reports and analyses were kept under wraps, as was the fact that the government was taking the saucers seriously.[14]

The U.S. Air Force's ongoing UFO investigation (Project Sign, Project Grudge, and Project Blue Book) collected more than twelve thousand reports, most of which were "explained." It was official policy to refuse to comment on "unexplained" cases. By keeping case details secret, the public was kept from learning that many of the allegedly explained cases had not been analyzed by generally accepted scientific standards.[15]

---

[13] SAC Memo to FBI, "Protection of Vital Installations," January 31, 1949.

[14] Smith, Wilbert, Memo to the Department of Transport, Ottawa, November 21, 1950.

[15] Bolender, Brig. Gen. C.H., ibid.

In 1976, with the amendment of the Freedom of Information and Privacy Acts by the U.S. Congress, a mechanism
was created for unearthing government UFO information
whose very existence had long been denied. Formal requests,
followed by appeals and sometimes legal action, produced
thousands of pages of previously classified documents from
the Air Force, Central Intelligence Agency, Federal Bureau
of Investigation, and other intelligence-oriented agencies.

It appears, however, that the released information was
the *least sensitive* material in the official files. Almost all the
released documents had been classified merely "confidential" or "secret," with just a few having been "top secret."
Many pages of these documents showed the black marks of
censorship. In fact, many pages of the voluminous case files
of the official U.S. Air Force investigation contained black
marks hiding information.[16]

The rapid flow of UFO documents in the 1970s dropped
to a slow trickle in the 1980s, but picked up again with the
administration's recent declassification measures. However,
since every government agency has at its disposal a long list
of reasons for refusing to release information, it will still be
easy to keep the most interesting and significant material
locked up.

The most striking example of continuing government secrecy is its reaction to growing public and press interest in
the apparent crash in 1947 of a strange craft on a sheep ranch
in New Mexico: the so-called Roswell Incident.[17] Most of
the time since 1947, the Air Force claimed that the crash was
that of a weather balloon. Despite the testimony to the contrary of dozens of first- and secondhand witnesses to this
event, the U.S. government has *yet* to release even one Air
Force report that includes the *full* testimony of these witnesses. Personal efforts in 1993 by U.S. Congressman Steven

---

[16] Judge Gesell Ruling, ibid.

[17] FBI teletype, July 5, 1947.

Schiff from New Mexico to learn about the crash were ignored. He turned the task over to the General Accounting Office, the investigative arm of the U.S. Congress.[18]

As a result of this investigation, the U.S. Air Force issued a brief report in July 1994 and a large report in 1995, both of them *now* stating that the wreckage found on the sheep ranch was *not* that of a balloon used for weather-data collection, but of a balloon from a then-secret Project Mogul experiment intended to detect Soviet nuclear explosions, which used *trains* and *clusters* of standard weather balloons.[19]

The GAO, in its final report in July 1995, stated that it could find no evidence for a UFO wreckage, *but* discovered that a large quantity of potentially valuable U.S. Air Force message traffic for the period had been *improperly destroyed*. Furthermore, since *no documentation* was found to support the new Project Mogul explanation, the GAO did *not* endorse the current Air Force explanation and stated that "the debate on what crashed at Roswell continues."[20]

While there is some indication that a few governments are easing their long-held policies of withholding all UFO information, there is no sign that this could become a trend or that it could produce truly meaningful information.

As a result of long-term and highly effective practices by many of the world's governments, the people have been kept in the dark about the extent and significance of UFO activity. Moreover, thousands of talented scientists who might

---

[18] Claiborne, William, "GAO Turns to Alien Turf in Probe," *Washington Post,* January 14, 1994.

[19] Weaver, Col. Richard L., USAF, "Report of Air Force Research regarding the 'Roswell Incident,' " July 1994. USAF, "The Roswell Report: Fact versus Fiction in the New Mexico Desert," October 1995.

[20] United States General Accounting Office Report to the Honorable Steven H. Schiff, House of Representatives. Government Records "Results of a Search for Records Concerning the 1947 Crash Near Roswell, New Mexico," July 1995.

contribute to the understanding of UFOs have been prevented from doing so because they are not part of the governmental system.

Since *no* government has *openly stated* that UFOs constitute a potential security threat, there is no reason to assume that there is any reasonable basis for continuing to keep UFO-related information secret.

SUMMARY OF QUOTATIONS

UFOs: The Reality

General Nathan D. Twining, Chairman of the Joint Chiefs of Staff (1957–1960):

> *The phenomena reported is something real and not visionary or fictitious . . . There are objects probably approximating the shape of a disc, of such appreciable size as to appear to be as large as a man-made aircraft . . . The reported operating characteristics such as extreme rates of climb, maneuverability (particularly in roll), and action which must be considered evasive when sighted or contacted by friendly aircraft and radar, lend belief to the possibility that some of the objects are controlled either manually, automatically, or remotely.* [Letter to the Commanding General of the U.S. Army Air Forces, September 23, 1947.]

Brigadier General João Adil Oliveira, Chief of the Air Force General Staff Information Service, and Director of the first official military UFO inquiry in Brazil in the mid-1950s:

> *It is impossible to deny any more the existence of flying saucers at the present time . . . The flying saucer is not a ghost from another dimension or a mysterious dragon. It is a fact confirmed by material evidence. There are thousands of documents, photos,*

and sighting reports demonstrating its existence.
["How to doubt?," *O Globo,* Rio de Janeiro, February 28, 1958.]

General Lionel M. Chassin, Commanding General of the French Air Forces, and General Air Defense Coordinator, Allied Air Forces, Central Europe (NATO):

> *The number of thoughtful, intelligent, educated people in full possession of their faculties who have "seen something" and described it grows every day ... We can ... say categorically that mysterious objects have indeed appeared and continue to appear in the sky that surrounds us ... [they] unmistakably suggest a systematic aerial exploration and cannot be the result of chance. It indicates purposive and intelligent action.* [Chassin, L., Foreword to the book by Michel Aime, *Flying Saucers and the Straight Line Mystery,* New York: Criterion Books, 1958.]

Admiral Roscoe Hillenkoetter, first Director of the CIA (1947–1950):

> *Unknown objects are operating under intelligent control ... It is imperative that we learn where UFOs come from and what their purpose is.* [Maccabee, Bruce, "What The Admiral Knew: UFO, MJ-12 and R. Hillenkoetter," *International UFO Reporter,* Nov./Dec., 1986.]

UFOs: Extraterrestrial Origin

Professor Hermann Oberth, German rocket expert considered one of the three fathers of the space age. In 1955, Dr.

Werner von Braun invited him to the United States, where he worked on rockets with the Army Ballistic Missile Agency and later with NASA:

> *It is my thesis that flying saucers are real and that they are space ships from another solar system. I think that they possibly are manned by intelligent observers who are members of a race that may have been investigating our earth for centuries.* [Oberth, H., "Flying Saucers Come From A Distant World," *The American Weekly,* October 24, 1954.]

General Kanshi Ishikawa, Chief of Staff of Japan's Air Self-Defense Force; Commander of the 2nd Air Wing, Chitose Air Base (1967):

> *Much evidence tells us UFOs have been tracked by radar; so, UFOs are real and they may come from outer space . . . UFO photographs and various materials show scientifically that there are more advanced people piloting the saucers and motherships.* [1967 interview published in *UFO News*, vol. 6, no. 1, 1974.]

Gordon Cooper, Astronaut (Mercury-Atlas 9, Gemini 5), Col., USAF (Ret.):

> *I believe that these extra-terrestrial vehicles and their crews are visiting this planet from other planets, which obviously are a little more technically advanced than we are here on earth. I feel that we need to have a top level, coordinated program to scientifically collect and analyze data from all over the earth concerning any type of encounter, and to determine how best to interface with these visitors in a friendly fashion.* [Letter to Grenada's Ambassador to the United Nations, November 9, 1978.]

Major-General Pavel Popovich, pioneer Cosmonaut and "Hero of the Soviet Union," President of All-Union Ufology Association of the Commonwealth of Independent States:

> Today it can be stated with a high degree of confi-
> dence that observed manifestations of UFOs are no
> longer confined to the modern picture of the world . . .
> The historical evidence of the phenomenon . . . allows
> us to hypothesize that ever since mankind has been
> co-existing with this extraordinary substance, it has
> manifested a high level of intelligence and tech-
> nology. The UFO sightings have become the con-
> stant component of human activity and require a
> serious global study . . . The scientific study of the
> UFO phenomenon should take place in the midst of
> other sciences dealing with man and the world.
> [Popovich, P., *MUFON 1992 International Symposium
> Proceedings*.]

UFOs: Secrecy and National Security

Wilbert Smith, Senior radio engineer, Department of Transport, Director of Project Magnet, the first Canadian government UFO investigation in the 1950s:

> The matter is the most highly classified subject in the
> United States Government, rating higher even than
> the H-bomb. Flying saucers exist. Their modus oper-
> andi is unknown but a concentrated effort is being
> made by a small group headed by Doctor Vannevar
> Bush. The entire matter is considered by the United
> States authorities to be of tremendous significance.
> [Top-secret memorandum on "Geo-Magnetics," No-
> vember 21, 1950.]

Dr. Paul Santorini, Greek physicist and engineer credited with developing the proximity fuse for the Hiroshima atomic bomb, two patents for the guidance system used in the U.S. Nike missiles, and a centrimetric radar system. In 1947, he investigated a series of UFO reports over Greece that were initially thought to be Soviet missiles:

> *We soon established that they were not missiles . . . Foreign scientists flew to Greece for secret talks with me . . . A world blanket of secrecy surrounded the UFO question because the authorities were unwilling to admit the existence of a force against which we had no possibility of defense.* [Fowler, R., *UFOs: Interplanetary Visitors,* 1974.]

Senator Barry M. Goldwater, Sr. (R-Arizona), Republican presidential candidate 1964:

> *The subject of UFOs is one that has interested me for some long time. About ten or twelve years ago, I made an effort to find out what was in the building at Wright Patterson Air Force Base where the information is stored that has been collected by the Air Force, and I was understandably denied the request. It is still classified above Top Secret.* [Good, T., *Above Top Secret,* Quill William Morrow, 1988; Frontispiece, letter to Shlomo Arnon, March 28, 1975.]

Representative Steven H. Schiff (R-New Mexico), in response to inquiries in 1993 concerning a possible cover-up of the crash of an alleged UFO outside Roswell, New Mexico, in 1947, requested information from the Department of Defense:

> *It's difficult for me to understand, even if there was a legitimate security concern in 1947, that it would be*

a present security concern these many years later. Frankly I am baffled by the lack of responsiveness on the part of the Defense Dept. on this one issue, I simply can't explain it. [Remarks on CBS radio's *The Gil Gross Show*, February 1994.]

UFOs: Challenge for Today's Science

Dr. J. Allen Hynek, Chairman of the Department of Astronomy at Northwestern University and scientific consultant to the U.S. Air Force investigations of UFOs from 1948 until 1969 (Projects Sign, Grudge, and Blue Book):

There exists a phenomenon . . . that is worthy of systematic rigorous study . . . The body of data point to an aspect or domain of the natural world not yet explored by science . . . When the long awaited solution to the UFO problem comes, I believe that it will prove to be not merely the next small step in the march of science but a mighty and totally unexpected quantum jump. [Hynek, J. Allen, *The UFO Experience: A Scientific Inquiry,* Chicago: Regnery Co., 1972.]

Dr. Felix Y. Zigel, Professor of Mathematics and Astronomy at the Moscow Aviation Institute, father of Russian Ufology:

The important thing now is for us to discard any preconceived notions about UFOs and to organize on a global scale a calm, sensation-free and strictly scientific study of this strange phenomenon. The subject and aims of the investigation are so serious that they justify all efforts. It goes without saying that international cooperation is vital. [Zigel, F., "Unidentified Flying Objects," *Soviet Life,* no. 2 (137), February 1968.]

M. Robert Galley, French Minister of Defense (1974):

I believe that the attitude of spirit that we must adopt vis-a-vis this phenomena is an open one, that is to say that it doesn't consist in denying apriori, as our ancestors of previous centuries did deny many things that seem nowadays perfectly elementary. [Bourret, Jean-Claude, *La nouvelle vague des soucoups volantes,* Paris: editions france-empire, 1975.]

Dr. Peter A. Sturrock, Professor of Space Science and Astrophysics and Deputy Director of the Center for Space Sciences and Astrophysics at Stanford University:

The definitive resolution of the UFO enigma will not come about unless and until the problem is subjected to open and extensive scientific study by the normal procedures of established science. This requires a change in attitude primarily on the part of scientists and administrators in universities. [Sturrock, Peter A., *Report on a Survey of the American Astronomical Society concerning the UFO Phenomenon,* Stanford University Report SUIPR 68IR, 1977.]

UFOs: The Effect of Ridicule

Admiral Roscoe Hillenkoetter (see page 34):

It is time for the truth to be brought out in open Congressional hearings. Behind the scenes high ranking Air Force officers are soberly concerned about the UFOs. But through official secrecy and ridicule, many citizens are led to believe the unknown flying objects are nonsense. [Statement in a NICAP news release, February 27, 1960.]

Dr. Frank B. Salisbury, Professor of Plant Physiology at Utah State University:

> *I must admit that any favorable mention of the flying saucers by a scientist amounts to extreme heresy and places the one making the statement in danger of excommunication by the scientific theocracy. Nevertheless, in recent years I have investigated the story of the unidentified flying object (UFO), and I am no longer able to dismiss the idea lightly.* [Paper on "Exobiology," presented at the First Annual Rocky Mountain Bioengineering Symposium, May 1964. Quoted in Fuller, John G., *Incident at Exeter,* Putnam, 1966.]

Representative Jerry L. Pettis (R-California) stated in 1968 during the House Committee on Science and Astronautics UFO hearings:

> *Having spent a great deal of my life in the air, as a pilot . . . I know that many pilots . . . have seen phenomena that they could not explain. These men, most of whom have talked to me, have been very reticent to talk about this publicly, because of the ridicule that they were afraid would be heaped upon them . . . However, there is a phenomena here that isn't explained.* [U.S. House of Representatives, Ninetieth Congress, July 1968.]

Dr. Peter A. Sturrock (see page 39):

> *In their public statements (but not necessarily in their private statements), scientists express a generally negative attitude towards the UFO problem, and it is interesting to try to understand this attitude. Most scientists have never had the occasion to confront evidence concerning the UFO phenomenon. To a scientist, the main source of hard information (other*

than his own experiments' observations) is provided by the scientific journals. With rare exceptions, scientific journals do not publish reports of UFO observations. The decision not to publish is made by the editor acting on the advice of reviewers. This process is self-reinforcing: the apparent lack of data confirms the view that there is nothing to the UFO phenomenon, and this view works against the presentation of relevant data. [Sturrock, Peter A., *Journal of Scientific Exploration,* vol. 1, no. 1, 1987.]

PART 2
CASE HISTORIES

INTRODUCTION

UFOs, under one name or another, have been described throughout history in ancient texts and paintings (see the photo insert.). Researchers place the beginning of the modern UFO era in the mid-1940s. Since then, strange sightings have been reported by tens of thousands of people from all parts of the world. Officials have exerted great effort to convince the public that UFOs have no validity and that they are no more than mistaken observations of natural phenomena and man-made objects.

To be sure, most sightings of UFOs can be explained as honestly mistaken responses to bright stars and planets, unusual clouds, unfamiliar airplanes, balloons, and satellites. These are known in UFO literature as Identified Flying Objects, or IFOs. Most cases *are* IFOs, but not *all*. A large number of credible UFO reports have been triggered by the appearance of "manufactured devices" that cannot be tied to known aircraft or spacecraft, after thorough analysis by competent investigators. Because of their appearance and/or behavior, they fall well outside the limits of known technology.

There is hardly a single country that has not experienced sightings in the past fifty years, most of which were never reported in UFO literature. The world's largest *nongovernmental* collection of UFO sightings that have been reported (UFOCAT) includes more than 50,000 cases. This total far exceeds the 12,500 reports in the Project Blue Book files. Of the UFOCAT cases, there are approximately 14,240 from Europe, 4,160 from South America, 4,300 from Oceania, 735 from Africa, and 27,450 from North America. The most active European countries include Great Britain with almost 7,000, France with 2,320, Germany with 1,260, and Spain with 1,200. Australia has had 3,220. In South America, the most active countries have been Argentina with 1,425 and Brazil with 1,125. Even Antarctica has had almost 50 reports.[21]

The case for UFO reality rests on the accumulation of reports that cannot be explained as "normal phenomena." Because of similar characteristics of appearance (shape, details) and/or behavior (maneuverability, speed, silence, etc.), they cannot be correlated with anything familiar and therefore must be placed in a separate category. These cases are represented in the Project Blue Book files by the approximately 600 officially unexplained cases. However, other investigating organizations report much higher numbers of unexplained cases.

Since the presentation of hundreds of cases would be impractical, a few prime examples that are particularly well documented and that reveal particular characteristics are in order.

---

[21] UFOCAT, a computerized catalog maintained by the Center for UFO Studies, Chicago, Illinois. By the end of 1993, it included 50,939 reports.

1944–45: FOO FIGHTERS OVER EUROPE AND ASIA

Although reports of sightings that were eventually termed "UFOs" can be traced far back into history, students of the subject have arbitrarily placed the beginning of the modern era in the mid-1940s with the appearance of UFOs over both the European and Pacific theaters of war. These UFOs were called by many names, all of which revealed a lack of understanding of their nature and source. To the Allies, they were "kraut fireballs" or "foo fighters," with the latter term surviving. It is believed that the Germans and Japanese saw them also.

Reports of "unexplained transparent, metallic and glowing balls" began in quantity in June 1944, at about the same time as the Allies invaded France, and Nazi Germany began launching V-1 flying bombs aimed at London, thus starting the era of unmanned missiles. Reports intensified in November 1944, not long after the first German V-2 ballistic rockets were fired at London and Paris.

Pilots and their crews reported that the "odd things" flew in formation with their airplanes, "played tag" with them, and generally behaved as if they were under intelligent control. At no time were they said to have displayed aggressive behavior. Nevertheless, most people assumed they were an experimental enemy device being prepared for operational use. Rumors of highly advanced weapons were common at that time, fed by the awesome reality of the V-1 and V-2 weapons. The following are typical of the scores of foo-fighter reports on record. Rumors persist that the U.S. Eighth Air Force in England commissioned a study on these reports, but no documentary evidence has yet been found.

On August 10, 1944, over the Indian Ocean, the co-pilot of a U.S. Army Air Force B-29 Superfortress heavy bomber reported:

*A strange object was pacing us about 500 yards [475
m.] off the starboard wing. At that distance it appeared as a spherical object, probably five or six feet
[1½–2 m.] in diameter, of a very bright and intense
red or orange . . . it seemed to have a halo effect.*

*My gunner reported it coming in from about a
5 o'clock position (right rear) at our level. It seemed
to throb or vibrate constantly. Assuming it was some
kind of radio-controlled object sent to pace us, I went
into evasive action, changing direction constantly, as
much as 90 degrees and altitude of about 2,000 feet
[600 m.]. It followed our every maneuver for about
eight minutes, always holding a position about 500
yards [475 m.] out and about 2 o'clock (right front)
in relation to the plane. When it left, it made an
abrupt 90 degree turn, accelerating rapidly, and disappeared into the overcast.[22]*

On December 22, 1944, over Hagenau, Germany, the
pilot and radar operator of an American night fighter encountered two "large orange glows" that climbed rapidly
toward them. When the pilot dove steeply and banked
sharply, the objects stayed with him. The pilot stated:

*Upon reaching our altitude, they levelled off and
stayed on my tail . . . After two minutes, they peeled
off and turned away, flying under perfect control.[23]*

Documents regarding foo-fighter incidents are still being
discovered even fifty years after the end of World War II. In
1992, researcher Barry Greenwood of Citizens Against UFO
Secrecy (CAUS) went to the National Archives in Suitland,

---

[22] Clark, Jerome, and Farish, Lucius, "The Mysterious 'Foo Fighters'
of World War II," *1977 UFO Annual.*

[23] Ibid.

Maryland, and located fifteen "Mission Reports" from the 415th Night Fighter Squadron, covering a period between September 1944 and April 1945. Here are two samples:

> *December 22/23, 1944—Mission 1, 17:05–18:50. Put on bogie by Blunder at 17:50 hours, had A.I. [Airborne Intercept radar] contact 4 miles range at Q-7372. Overshot and could not pick up contact again. A.I. went out and weather started closing in so returned to base. Observed 2 lights, one of which seemed to be going on and off at Q-2422.*

> *February 13/14, 1945—Mission 2, 18:00–20:00. About 19:10, between Rastatt and Bishwiller, encountered lights at 3,000 feet, two sets of them, turned into them, one set went out and the other went straight up 2–3,000 feet [600–900 m.], then went out. Turned back to base and looked back and saw lights in their original position again.[24]*

Suggested explanations, both at the time and subsequently, have included prototype enemy antiaircraft devices, St. Elmo's fire (glowing balls of static electricity), and simple misidentification of other airplanes.[25]

In order to accept any of the above explanations, one would have to discount the observational skills of scores of veteran combat pilots and their crew members whose very survival depended on their ability to instantly identify and react to any potential threat.

---

[24] Greenwood, Barry, "More Foo-Fighter Records Released," *Just Cause,* no. 33, CAUS, September 1992.

[25] Chamberlain, Jo, "The Foo Fighter Mystery," *The American Legion Magazine,* December 1965; Associated Press article, "Nazi Fire Balls May Be Kind of Ball Lightning," *New York Herald Tribune,* January 3, 1945; other, miscellaneous press reports.

1946: GHOST ROCKETS
OVER SCANDINAVIA

Barely a year after the foo-fighter episodes, the second wave of UFO sightings began, this time in Scandinavia.

On the night of June 9, 1946, a brilliant light streaked over Helsinki, Finland, with a smoke trail and the sound of thunder; its luminous trail persisted for ten minutes. Had this not been repeated the next night, it would have been written off as an unusually large meteor. The second one, according to news reports, turned and went back in the direction from which it had come.

On June 12, the Swedish Defense Staff asked military personnel to report their sightings through official channels, admitting that they had been aware of the phenomena since May. On July 9 alone, more than two hundred reports were received, many of them describing tubular or "spindle-shaped" objects flying low and slowly, with little or no sound.

A week after the establishment of a special "ghost rocket" committee by the Swedish government, American Secretary of the Navy, James Forrestal, traveled to Stockholm to meet with the Swedish Secretary of War. According to a secret FBI memo of August 19, 1947, "the 'high brass' of the War Department exerted tremendous pressure on the (Army) Air Force's Intelligence to conduct research and collect information in an effort to identify the sightings."[26]

On August 11, 1946, more than three hundred reports of strange sightings were observed in just the Stockholm area. On August 20, General Jimmy Doolittle (in Stockholm on business for the Shell Oil Company) met with the head of the Swedish Air Force. This led to wide speculation in the Swedish press, as well as *The New York Times*, that "ghost

---

[26] FBI Report on "Flying Discs," August 19, 1947.

rockets" were the subject of the meeting. In the 1980s, however, in an interview with UFO researchers, General Doolittle denied that his Swedish trip was officially connected with the ghost rockets, although it is certainly likely that the subject came up in casual conversation.[27]

Soon thereafter, Swedish newspapers began censoring most reports of ghost rockets. However, reports appeared in other Scandinavian countries. According to a British Air Ministry Intelligence Report of September 1946:

> *A large number of visual observations have been obtained from Scandinavia. Some of the best came from Norway. An analysis suggests the most notable characteristics of the projectiles to be: a) great speed; b) intense light frequently associated with missile; c) lack of sound; d) approximate horizontal flight . . . Thus, if the phenomena now observed are of natural origin, they are unusual; sufficiently unusual to make possible the alternative explanation that at least some are missiles. If this is so, they must be of Russian origin.[28]*

There was a concerted effort on the part of the Swedish government to blame many of the sightings on Soviet tests of captured German rockets. The Soviet Union had occupied Peenemünde, the secret German test site across the Baltic Sea, where the V-1 and V-2 missiles were developed. Years later it was learned that the captured German equipment was immediately moved to Poland. There were no Soviet tests at Peenemünde, and thus the "official" explanation for the ghost rockets proved impossible.

---

[27] Liljegren, Anders, "General Doolittle and the Ghost Rockets," *AFU Newsletter,* no. 36, Jan./Dec. 1991. Archives for UFO Research, Sweden.

[28] British Air Ministry Report, "Investigation of Reported Missile Activity Over Scandinavia," September 9, 1946.

As reports from Scandinavia began to taper off in September 1946, they were replaced by reports of similar sightings from Hungary, Greece, Morocco, and Portugal. In 1984, when the Swedish government finally opened its ghost-rocket files, researchers found that more than fifteen hundred reports had been secretly collected from 1946 on. One of the few official American reactions to the ghost rockets came in the January 9, 1947, issue of the Defense Department's *Intelligence Review* (classified "secret" until 1978). This four-page summary of the ghost-rocket events suggests that some of the sightings may have been of Soviet test missiles or jet airplanes (although no jets are known to have been in or near Scandinavia at the time).[29]

One sighting, detailed in the FBI report cited on page 50, suggests there may have been more to it:

> *On 14, August (1946) at 10 A.M. [a Swedish Air Force pilot] . . . was flying at 650 feet [200 m.] over central Sweden when he saw a dark, cigar-shaped object about 50 feet [15 m.] above and approximately 6,500 feet [2 km.] away from him travelling at an estimated 400 mph [650 km./hr.]. The missile had no visible wings, rudder or other projecting part; and there was no indication of any fuel exhaust (flame or light), as had been reported in the majority of other sightings.*
>
> *The missile was maintaining a constant altitude over the ground and, consequently, was following the large features of the terrain. This statement casts doubt on the reliability of the entire report because a missile, without wings, is unable to maintain a constant altitude over hilly terrain.*[30]

---

[29] *Intelligence Review,* number 49, January 9, 1947, "Ghost Rockets Over Scandinavia."

[30] FBI Report, ibid.

Many years later, sophisticated cruise missiles, with tiny wings that were invisible at such a distance, achieved "terrain-following" flight as a matter of routine. In 1946, this was far beyond the capability of any existing technology.

Perhaps the lingering mystery of the ghost rockets was best expressed by Air Engineer Eric Malmberg, once secretary of Sweden's Defence Staff committee on the matter, who was interviewed forty years later:

> *I would like to say that everyone on the committee, as well as the chairman himself, was sure that the observed phenomena didn't originate from the Soviet Union. Nothing pointed to that solution.*
>
> *On the other hand, if the observations are correct, many details suggest that it was some kind of a cruise missile that was fired on Sweden. But nobody had that kind of sophisticated technology in 1946.[31]*

[31] Liljegren, Anders, and Svahn, Clas, "Ghost Rockets and Phantom Aircraft," paper in the anthology *Phenomenon—Forty Years of Flying Saucers*, Avon Books, 1989.

1947:
FIRST AMERICAN SIGHTING WAVE

The first major wave of American sightings produced more than a thousand reports, the term *flying saucer*, and the first confirmed investigations by the U.S. government. The reports were from all the lower 48 states, mainly of round objects seen in the daytime. For two weeks, especially around the July Fourth weekend, newspapers and radio broadcasts were filled with stories of flying saucers and flying discs. Early official studies concluded that they were real and unexplained.

It began on the afternoon of June 24, 1947, with the sighting of a formation of strange high-speed objects. Kenneth Arnold, flying his single-engine Callair airplane over southwestern Washington State, had interrupted his business trip to assist in the search for a missing military transport plane. From the official U.S. Army Air Force report on the event:

> *I hadn't flown more than two or three minutes on my (new) course when a bright flash reflected on my airplane. It startled me as I thought I was too close to some other aircraft. I looked every place in the sky and couldn't find where the reflection had come from until I looked to the left and the north of Mt. Rainier where I observed a chain of nine peculiar looking aircraft flying from north to south at approximately 9,500 feet [3,000 m.] elevation and going, seemingly, in a definite direction of about 170 degrees.*
>
> *They were approaching Mt. Rainier very rapidly, and I merely assumed they were jet planes. Anyhow, I discovered that this was where the reflection had come from, as two or three of them every few seconds would dip or change their course slightly, just enough*

for the sun to strike them at an angle that reflected brightly at my plane.

I thought it was very peculiar that I couldn't find their tails but assumed they were some kind of jet plane. I was determined to clock their speed, as I had two definite points I could clock them by; the air was so clear that it was very easy to see objects and determine their approximate shape and size at almost 50 miles [80 km.] that day.

[The clock] . . . on my instrument panel, read one minute to 3 P.M. as the first object of this formation passed the southern edge of Mt. Rainier . . . I would estimate their elevation could have varied a thousand feet [300 m.], one way or another, up or down, but they were pretty much on the horizon to me, which would indicate they were near the same elevation as I was.

They seemed to hold a definite direction but rather swerved in and out of the high mountain peaks. Their speed at the time did not impress me particularly, because I knew that our army and air forces had planes that went very fast.

What kept bothering me as I watched them flip and flash in the sun right along their path was the fact that I couldn't make out any tail on them, and I am sure that any pilot would justify more than a second look at such a plane.

I observed them quite plainly, and I estimated my distance from them, which was almost at right angles, to be between 20 and 25 miles [30–40 km.]. I knew they must be very large to observe their shape at the distance, even on as clear a day as it was that Tuesday. In fact I compared a . . . fastener or cowling tool I had in my pocket with them—holding it up on them and holding it up on the DC-4 (airliner)—that I could observe at quite a distance to my left, and they seemed smaller than the DC-4; but, I should judge their span would have been as wide as the furthest engines on each side of the fuselage of the DC-4. [Note: this span is about 55 ft. or 16 m.]

I could quite accurately determine their pathway due to the fact there were several high peaks that were a little this side of them as well as higher peaks on the other side of their pathway.

As the last unit of this formation passed the southernmost high snow-covered crest of Mt. Adams, I looked at my sweep second hand and it showed that they had travelled the distance in one minute and 42 seconds. Even at the time, this timing did not upset me as I felt confident that after I landed there would be some explanation of what I had seen. [Note: 48 miles in 1:42 seconds works out to 1,700 mph or 2,700 km./hr., at a time when the official World Speed Record was 624 mph or 1,000 km./hr.]

A number of newsmen and experts suggested that I might have been seeing reflections or even a mirage. This I know to be absolutely false, as I observed these objects not only through the glass of my airplane but turned by my airplane sideways where I could open my window and observe them with a completely un-obstructed view, without sun glasses . . . They seemed longer than wide, their thickness was about 1/20th of their width.[32]

To accompany his statement, Arnold added a simple sketch of one of the objects: a circle with the rear flattened or clipped off.

It was not known at the time, but others had seen forma-tions of strange objects in the Pacific Northwest on the same day. That morning, five or six discs were seen banking and

---

[32] Arnold, Kenneth, report to the U.S. Army Air Force, June 1947. Reprinted in Steiger, Brad, ed., *Project Blue Book,* Ballantine Books, 1976.

circling from the same Cascade Mountain Range over which Arnold had been flying during his sighting; at 2:30 P.M., three flat discs were seen tilting as they flew from Richland, Washington, 100 miles (160 km.) to the east; at 3:00 P.M., a man saw nine discs in formation from Mineral, Washington, almost directly beneath Arnold's airplane.

The conclusion drawn later by Project Blue Book was that Arnold had failed to identify some conventional airplanes. No specifics were suggested, as there were no circular, disc, or similarly shaped airplanes flying in or near the United States, nor were there any airplanes in the world capable of even half the speed at which he clocked the formation. The other sightings of formations in the same area do appear in newspaper reports, but not in the official files.

This marked the beginning of the 1947 UFO sighting wave. A study of newspapers by Ted Bloecher lists 832 sighting reports between June 15 and July 15. An expansion of his study raised the total to approximately 1,500 separate reports, including many from outside the United States.[33]

The files of Project Blue Book show barely fifty reports for the period. Nevertheless, the first known official study of the 1947 sighting wave concluded, from just thirteen of those reports:

From detailed study of reports selected for their impression of veracity and reliability, several conclusions have been formed:

(a) This "flying saucer" situation is not all imaginary or seeing too much in some natural phenomena. Something is really flying around.

(b) Lack of topside inquiries, when compared to the prompt and demanding inquiries that have originated

---

[33] Bloecher, Ted, *Report on the UFO Wave of 1947*, NICAP, 1968.

*topside upon former events, give more than ordinary
weight to the possibility that this is a domestic project
about which the President, etc., know.*

*Whatever the objects are, this much can be said of
their physical appearance:*

*1. The surface of these objects is metallic, indicating
a metallic skin, at least.*

*2. When a trail is observed, it is lightly colored, a
blue-brown haze, that is similar to a rocket engine's
exhaust. Contrary to a rocket of the solid type, one
observation indicates that the fuel may be throttled
which would indicate a liquid rocket engine.*

*3. As to shape, all observations state that the object is
circular or at least elliptical, flat on the bottom and
slightly domed on the top. The size estimates place it
somewhere near the size of a C-54 or a Constellation.
[Note: 1940s airliners had a wingspan of 120 ft. or
35 m., and length of 95 ft. or 30 m.]*

*4. Some reports describe two tabs, located at the rear
and symmetrical about the axis of flight motion.*

*5. Flights have been reported, from three to nine
of them, flying good formation on each other, with
speeds always above 300 kts. [350 mph or 650
km/hr.].*

*6. The discs oscillate laterally while flying along,
which could be snaking.*[34]

---

[34] Study by Air Force Base Intelligence Report, "Flying Discs," ibid.

Official interest in the phenomena was demonstrated in a famous September 1947 memo from General Nathan D. Twining, Chief of the Air Materiel Command at Wright Field, Ohio:

> *a. The phenomena reported is something real and not visionary or fictitious.*
>
> *b. There are objects probably approximating the shape of a disc, of such appreciable size as to appear to be as large as a man-made aircraft.*
>
> *c. There is a possibility that some of the incidents may be caused by natural phenomena, such as meteors.*
>
> *d. The reported operating characteristics such as extreme rates of climb, maneuverability (particularly in roll), and action which must be considered evasive when sighted or contacted by friendly aircraft and radar, lend belief to the possibility that some of the objects are controlled either manually, automatically, or remotely.*[35]

Following other studies and reports, the U.S. Air Force instituted its first announced UFO investigation in January 1948: Project Sign.

---

[35] Twining, Gen. Nathan, Memo to Commanding General Army Air Forces re "AMC Opinion Concerning 'Flying Discs,' " September 23, 1947.

1952: SECOND AMERICAN SIGHTING WAVE

From 1947 through 1951, the U.S. Air Force UFO investigation (at first Project Sign, then Project Grudge, and by 1952 Project Blue Book) had logged 700 UFO reports, an average of just over 150 per year. The staff had yet to encounter much pressure, or much attention from the press or the public. It was a fairly routine military intelligence-gathering effort.[36]

This was the situation until the middle of 1952. The year started out as the preceding one had begun, with fewer than one sighting per day in the first three months. In April and May, the flow increased to three per day, with the rate doubling in June. For the first half of the year, there had been 300 reports, at four times the annual rate, and still the peak had not been approached.

For the first three weeks of July, there was an average of eight reports per day, many of them coming from Air Force jet interceptor pilots sent aloft in response to radar or visual sightings from the ground. Starting on the 22nd and lasting through the 29th, reports jumped to an average of 27 per day. By the end of that extremely busy month, almost 400 reports had been recorded, which was more than in any previous full year.

The Project Blue Book office at Wright Field, Dayton, Ohio, was headed by Captain Edward Ruppelt, whose tiny staff was completely overwhelmed by the volume of work. Reports poured in by mail, teletype, telephone, and messenger faster than they could be processed, let alone investigated. They were stacked up with vague plans to investigate when things finally calmed down.

---

[36] Ruppelt, Edward J., *The Report on Unidentified Flying Objects,* Doubleday & Co., 1956.

As important as the sheer number of reports received was the particular nature of some of them, especially those from three nights of intense activity over Washington, D.C. On July 19/20, July 26/27, and August 2/3, the skies above the nation's capital were crowded with UFOs darting here and there, over the White House, over the Capitol Building, over the Pentagon.

They were seen from the ground and from control towers at Washington National Airport, Bolling Air Force Base across the Potomac River, and from nearby Andrews Air Force Base. They were also tracked on radar from all three airfields, as radar operators conferred by telephone to ensure they were tracking the same targets. In many instances, airline pilots flying in the area were able to provide visual confirmation of radar tracking.

The appearance of unidentified objects flying with impunity over the heart of the American government and its military establishment was embarrassing to the Department of Defense, whose responsibility it was to protect the country from airborne intrusion. A flood of questions from reporters led the U.S. Air Force to call its biggest press conference since World War II.

It was held in Room 3E-369 of the Pentagon and was presided over by Air Force Intelligence Chief, Major General John Samford. The main explanation given for the rash of sightings over Washington was something called a "temperature inversion," which is the immediate cause of a mirage. General Samford suggested that lights on the ground may have looked like they were in the air because an inversion can act like an "air lens" and bend light rays. He added that something similar could have "tricked" radar into thinking it was tracking aerial targets, which were actually ground objects.[37]

The press left the ninety-minute conference confused, but convinced that the UFOs were no more than atmospheric

---

[37] Transcript of General Samford's press conference at the Pentagon, July 29, 1952.

phenomena. It wasn't until 1969 that an Air Force scientific report made it clear that inversions strong enough to create the effects with which General Samford credited them could not exist in the Earth's atmosphere! Moreover, probably no UFO report had ever been caused by a temperature inversion or mirage.[38]

The same day as General Samford held his press conference, the wheels began to turn at the Central Intelligence Agency. A memo from Ralph Clark, Acting Assistant Director for Scientific Intelligence, to the Deputy Director for Intelligence stated:

> *In the past several weeks, a number of radar and visual sightings of unidentified aerial objects have been reported. Although this office has maintained a continuing review of such reputed sightings during the past three years, a special study group has been formed to review this subject to date.*[39]

A few days later, a note was sent to Mr. Clark by Edward Tauss:

> *. . . so long as a series of reports remains "unexplainable" (interplanetary aspects and alien origin not being thoroughly excluded from consideration), caution requires that intelligence continue coverage of the subject.*[40]

CIA interest in the UFO phenomena increased and led to a secret panel of five prominent scientists, convened in Janu-

---

[38] Menkello, F.V., ibid.

[39] CIA memorandum to the Deputy Director/Intelligence, July 29, 1952, re "Recent Sightings of Unexplained Objects."

[40] Informal CIA memorandum to Deputy Assistant Director/SI, August 1, 1952, re "Flying Saucers."

ary 1953. The Scientific Panel on Unidentified Flying Objects was chaired by astrophysicist Dr. H. P. Robertson and included Dr. Luis Alvarez (who received the Nobel Prize for Physics many years later), Dr. Thornton Page of Johns Hopkins University and later NASA Johnson Space Center, and other top scientists. The negative conclusions of the so-called Robertson Panel would exert tremendous influence on all federal policy vis-a-vis UFOs. The panel recommended in part:

> *That the national security agencies take immediate steps to strip the unidentified flying objects of the special status they have unfortunately acquired.*[41]

---

[41] Central Intelligence Agency, "Report of the Scientific Panel on Unidentified Flying Objects," January 1953.

1956: RADAR/VISUAL
JET CHASE OVER ENGLAND

On the night of August 13–14, 1956, radar operators at two military bases in the east of England repeatedly tracked single and multiple objects that displayed high speed as well as rapid changes of speed and direction. Two jet interceptors were sent up and were able to see and track them in a brief series of maneuvers. According to official U.S. Air Force reports, the sightings could not be explained by radar malfunction or by unusual weather.[42]

It began at 9:30 P.M., when Airman 2nd Class John Vaccare, of the U.S. Air Force at RAF Bentwaters, tracked one UFO on his Ground Controlled Approach radar (type AN/MPN-11A) as it flew 40–50 miles (65 to 80 km.) in 30 seconds, i.e., 4,800 to 6,000 mph (7,500 to 9,500 km./hr.).

A few minutes later Vaccare reported to T/Sergeant L. Whenry that a group of twelve to fifteen unidentified targets was tracked from 8 miles (13 km.) southwest of Bentwaters to 40 miles (65 km.) northeast, at which time they "appeared to converge into one very large object, according to the size of the blip on the radar scope, which seemed to be several times larger than a B-36 aircraft [one of the largest operational bombers in history, with a wingspan of 230 feet or 70 m.]." The single large blip stopped twice for several minutes while being tracked, before flying off the scope.

At 10:00 P.M., a single unidentified target was tracked from Bentwaters as it covered 55 miles (90 km.) in just 16 seconds. This works out to over 12,000 mph (19,000 km./hr.).

Then, at 10:55 P.M., the Bentwaters GCA radar picked up an unidentified target on the same east-to-west course as the previous one, at an apparent speed of "2,000 to 4,000 mph"

---

[42] USAF Air Intelligence Information Report filed by Captain Edward L. Holt, August 31, 1956.

(3,200 to 6,400 km./hr.). Someone in the Bentwaters control tower reported seeing "a bright light passing over the field from east to west at about 4,000 feet [1,200 m.]." At about the same time, the pilot of a C-47 twin-engine military transport plane over Bentwaters said, "a bright light streaked under my aircraft travelling east to west at terrific speed." All three reports coincided.

Soon after, radars at Bentwaters and RAF Lakenheath reported a stationary object 20–25 miles (32–40 km.) southwest of the latter base. It suddenly began moving north at 400 to 600 mph (650 to 1,000 km./hr.), but "there was no build-up to this speed—it was constant from the second it started to move until it stopped." It made several abrupt changes of direction without appearing to slow for its turns.[43]

Around 11:30 P.M., the RAF launched a deHavilland Venom jet interceptor from RAF Waterbeach. According to the U.S. Air Force UFO report:

> *Pilot advised he had a bright white light in sight and would investigate. At 13 miles [20 km.] west he reported loss of target and white light. Lakenheath (radar) vectored him to a target 10 miles [16 km.] east of Lakenheath and pilot advised (that) target was on his radar and was "locking on." Pilot then reported he had lost target on his radar.*
>
> *Lakenheath GCA reports that as the Venom passed the target on radar, the target began a tail chase of the friendly fighter. Radar requested pilot acknowledge this chase. Pilot acknowledged and stated he would try to circle and get behind the target. Pilot advised he was unable to "shake" the target off his tail and requested assistance.*
>
> *One additional Venom was scrambled from RAF station. Original pilot stated: "Clearest target I have ever seen on radar."*

---

43 Ibid.

The following conversation between the two Venom fighter pilots was heard by the Lakenheath watch supervisor:

"Did you see anything?" [Pilot #2]

"I saw something, but I'll be damned if I know what it was." [Pilot #1]

"What happened?" [Pilot #2]

"He—or it—got behind me and I did everything I could to get behind him and I couldn't. It's the damndest thing I've ever seen." [Pilot #1][44]

The 1969 report by the Air Force–funded study at the University of Colorado under Dr. Edward U. Condon concluded:

In summary, this is the most puzzling and unusual case in the radar-visual files. The apparent rational, intelligent behavior of the UFO suggests a mechanical device of unknown origin as the most probable explanation of this sighting. However, in view of the inevitable fallibility of witnesses, more conventional explanations of this report cannot be entirely ruled out.[45]

---

[44] Ibid.

[45] Gillmor, Daniel S., ibid.

1957:
THIRD AMERICAN SIGHTING WAVE

The third of the major American waves of UFO reports peaked in the first week of November 1957, with at least 30 accounts of electrical devices experiencing temporary failure in connection with a UFO sighting.

The files of Project Blue Book show 330 reports for that week,[46] while the files of the private National Investigations Committee on Aerial Phenomena (NICAP) list almost 90 unexplained reports.

It started four weeks after the Soviet Union shocked the world by launching the first Earth-orbiting satellite, Sputnik I, and a day before Sputnik II was orbited with a small dog as passenger. Public enthusiasm for searching the night sky for a glimpse of the first satellite had waned, and that for the second had not yet begun.

The most striking feature of this sighting wave was the concentration of "electromagnetic effect" cases around the west Texas town of Levelland. There were at least eight such reports in the space of two and a half hours in an area to the west, north, and east of Levelland:

At 10:30 P.M. came the report from truck driver Pedro Saucedo, who described seeing a blue torpedo-shaped object with yellow flame and white smoke coming out of its rear. He estimated it was 200 feet [60 m.] long and 6 feet [2 m.] wide. He said it rose from a nearby field and roared low over his truck with a loud, explosive sound, and produced so much heat he got out of his truck and lay on the ground. "It sounded like thunder, and my truck rocked from the blast."

---

[46] Project Blue Book case files, U.S. National Archives, Washington, D.C.

*He thought it came within 200–300 feet [60–90 m.].
His truck lights and engine failed while the UFO was
in view; after it disappeared, his lights worked perfectly, and he was able to re-start the engine.*

*At Pettit, Texas, 10 miles [16 km.] to the north-
west, two grain combines failed as a UFO flew past.*

*Shortly before midnight, Jim Wheeler reported
seeing a large 200 ft. [60 m.] elliptical object on the
road; as he drove toward it, his car lights and engine
failed. The UFO rose and flew off, and when it blinked
off, his lights came back on and he was able to re-
start his engine.*

*At the same time, Jose Alvarez's car lights and en-
gine died when he saw a glowing, 200-foot [60 m.]
UFO nearby. After the object flew away, his lights came
back on and he was able to re-start his engine.*

*At about 12:05 A.M., college student Newell Wright's
car lights and engine failed. He got out to fix them,
looked up and saw a glowing, bluish-green, flat-
bottomed, oval object on the highway. The object was
in sight for four or five minutes. During that time,
Wright tried to start his engine, and while the starter
made contact, the motor was unaffected. The object
disappeared, straight up, and immediately the car
lights came back on, the engine started, and then op-
erated perfectly.*

*At 12:25 A.M., Frank Williams' car experienced
a failure of its lights and engine, when a glowing,
egg-shaped object appeared on or near the ground
pulsating brightly. When it rose straight up, the car
returned to normal. "When it took off, it sounded like
thunder."*

*At 12:45 A.M., Ronald Martin's truck lights and
engine stopped working when a round, glowing UFO
landed and changed from orange to blue-green. He
said the glow was so bright it lit up the inside of his
truck. The UFO then changed back to orange and*

took off straight up. The car lights came back on, and his engine re-started by itself!

At 1 A.M., some 17 miles [27 km.] to the north, Fire Marshall Ray Jones reported seeing a streak of light and at the same time his car lights dimmed and his engine almost quit.

At 1:15 A.M., James Long said he saw an elliptical UFO on the road ahead, and when he drove to within 200 feet [60 m.] of it, the lights and engine of his truck died. The UFO then shot up vertically with a sound like thunder, and the lights and engine returned to normal.

By 1:30 A.M., Hockley County Sheriff Weir Clem had heard so many reports that he decided to see for himself. He drove out with a deputy sheriff, and saw a large oval red light, though he did not experience electrical system problems. Years later he said: *"The object was shaped like a huge football and had bright white lights. The blinding lights flashed on, it went right over the car and was gone. No living human being could believe how fast it traveled. The whole thing was as bright as day; it lit up the whole area."*[47]

Project Blue Book sent a single investigator to Levelland to check the reports. His explanation, accepted as the official Air Force conclusion, was that:

. . . *the major cause for the Levelland case was a severe electrical storm. The storm stimulated the populace into a high level of excitement. This excitement reflected itself in their reactions to ordinary circumstances, and resulted in the inflation of the stories of some of the witnesses concerning their experiences.*[48]

---

[47] Webb, Walter N., NICAP Field Investigation Report, 1957.

[48] Project Blue Book report, "Levelland, Texas, November 2–4, 1957."

Ten years after these incidents, atmospheric physicist Dr. James McDonald completed a study and determined that there had been no storm in the area, and thus no source of excessive moisture to interfere with the automobiles' electrical systems. With no "severe electrical storm" to "stimulate the populace into a high level of excitement," the official explanation falls apart.[49]

---

[49] McDonald, James, "UFOs: Greatest Scientific Problem of Our Time?" lecture to the American Society of Newspaper Editors, Washington, D.C., April 22, 1967.

1958: BRAZILIAN NAVY PHOTOGRAPHIC CASE

On February 21, 1958, the Brazilian newspapers *Correio de Manha* and *O Jornal* published a sequence of clear, daytime photographs showing an oval object with a ring in the center, flying off the island of Trindade in the South Atlantic Ocean. The photos were taken by a professional civilian photographer, Almiro Barauna, on board the Brazilian Navy training ship *NE Almirante Saldanha*, which was conducting research for the International Geophysical Year (IGY). The Navy at first kept the matter secret, but the photos were eventually given to the press by the President of Brazil, Juscelino Kubitschek.[50] (See the photo insert.)

Other high-ranking officials, such as the Minister of the Navy, Admiral Alves Camera, were quoted in the press vouching for the photos. Following his weekly meeting with the President, the Minister told the United Press that "the Navy has a great secret which it cannot divulge because it cannot be explained."[51] On February 27, Deputy Sergio Magalhaes of the House of Representatives formally requested the Navy Ministry to answer several questions about the photos and other prior UFO observations at the IGY post, maintained by the Navy in Trindade Island.[52]

The Navy eventually released a detailed report on the matter titled "Clarification of the observation of unidentified

---

[50] Fontes, Olavo T., M.D., "UAO Sightings Over Trindade," originally published in three parts in *The A.P.R.O. Bulletin*, Alamogordo, New Mexico, January, March, and May 1960; reprinted in full as "The Brazilian Navy UFO Sighting At The Island Of Trindade," *Flying Saucers,* Amherst, Wisconsin, Feb. 1961.

[51] United Press News wire from Rio de Janeiro, February 25, 1958; reprinted in Fontes, O.T., ibid.

[52] Ibid.

flying objects sighted on the Island of Trindade, in the period of 12/5/57 to 1/16/58," prepared by Captain of Corvette (CC), Carlos Alberto Ferreira Bacellar, Commander of the Oceanographic Station at Trindade. The report begins with a short summary of five UFO sightings by Navy crewmen and workers at the station from early December 1957 to mid-January 1958, including a phenomenon seen by Captain Bacellar with a theodolite (a surveying instrument for measuring vertical and horizontal angles). Four observations occurred in daytime and one at night; one was considered likely to be a seagull.

The report then introduces Almiro Barauna as "a professional civilian photographer who was on deck in the stern of the ship, ready to photograph the operation of hoisting the launch," when he was "alerted about the UFO" and was able to take four photographs showing the object. The crucial details of the development of the film are discussed:

> *That, after having taken the above-mentioned photographs, the photographer, in the presence of CC Bacellar and other persons, took the roll of film from the camera; later, in the company of this official he went to the darkroom of the ship (improvised in the infirmary), dressed as he was in shirt and shorts, and where he remained only ten minutes, presenting at once the negative of the film to CC Bacellar, who affirms having seen the above-mentioned UFO represented on the negative, although with much less clarity because the film was somewhat dark.[53]*

---

[53] Six-page document from the Brazilian Department of the Navy, General Staff of the Fleet, Subdivision of Information, "SUBJECT: Clarification of the observation of unidentified flying objects sighted on the Island of Trindade, in the period of 12/5/57 to 1/16/58." English translation in the papers of the late Dr. Edward U. Condon at the American Philosophical Society in Philadelphia.

An analysis of the facts in the report confirms that there were many witnesses on deck of the *Almirante Saldanha* of "various qualifications—workmen, sailors, dentist, doctor, aviation officer and professional photographer," but no exact number is given. A report by Dr. Willy Smith, published by the Center for UFO Studies, indicates that "all in all, 48 ocular witnesses were on deck during the incident," although no source for this figure is provided. These included sailors, workers, and the ship's dentist, as well as members of a civilian submarine diving group to which Baruana and Brazilian Air Force Captain (Ret.) J. T. Viejas belonged.[54]

Captain Viejas's eyewitness description of the incident was published in the Brazilian press:

> *The first view was that of a disc shining with phosphorescent glow, which—even at daylight—appeared to be brighter than the moon. The object was about the apparent size (angular diameter) of the full moon. As it followed its path across the sky, changing to a tilted position, its real shape was clearly outlined against the sky: that of a flattened sphere encircled, at the equator, by a large ring or platform. Its speed was around 700 miles an hour [1,100 km./hr.] at the moment it disappeared into the horizon.*

Captain Viejas added that the sighting occurred at 12:20 P.M., causing "a tremendous confusion aboard. Mr. Baruana found it very difficult to operate his camera, being pushed and pulled by excited observers around him."[55] Neither CC Bacellar nor the Captain of the *Almirante Saldanha*, Jose Saldanha da Gama, observed the phenomenon, although they did see the commotion caused by the event.

---

[54] Smith, Willy, "Trindade Revisited," *International UFO Reporter,* CUFOS, July/August 1983.

[55] Fontes, O.T., ibid.

Baruana gave detailed interviews to Brazilian reporters. His camera was a Rolleiflex 2.8 camera, model E, "set at speed 125, with the aperture at f/8." He shot the first two photos before the object disappeared behind the peak "Desejado." The UFO then reappeared, "bigger in size and flying in the opposite direction, but lower and closer than before, and moving at a higher speed. I shot the third photo." The fourth and fifth photos were lost when Barauna was pushed by other witnesses. The last photo in the roll of film was taken when the object was moving back toward the sea.

Barauna also disclosed that he had been interrogated for four hours at the Navy Ministry and that:

> *Some days later I was called again. This time they [Navy] also asked for my Rolleiflex. They wanted to make tests in order to estimate, if possible, the speed of the flying saucer at the moment of the sighting. The tests were performed. They showed that I had taken my six pictures in 14 seconds, and that the saucer was flying at 900 to 1,000 km./hr. [550 to 600 mph].*[56]

The negatives were analyzed by Navy and civilian experts from the Cruzeiro do Sul Aerophotogrammetric Service. The previously cited Navy document states that a technician from the Hydrographic Navy Department concluded there were no signs of tampering with the negatives that showed "the object photographed." A "more complete and thorough examination" was made by photo technicians from Cruzeiro do Sul (a private airline company), "including microscopic, for the verification of granulation, verification of signs, luminosity, and details of contour." The Cruzeiro experts concluded:

---

[56] Interview with Almiro Barauna, *O Globo,* Rio de Janeiro, February 24, 1958; reprinted in Fontes, O.T., ibid.

There was on the above-mentioned negatives no sign of montage, all indicating it to be a negative of the object really photographed . . .

Any hypotheses of later montage were removed; it would be impossible to prove either the existence or nonexistence of prior montage, which requires, however, extreme technical skill and circumstances favorable to its execution.

The Navy's final conclusion was very cautious, due to Barauna's reputation as a highly skilled photographer with some experience in UFO montages, which he had previously shown while refuting the 1952 Barra de Tijuca UFO photo case. The report's two final conclusions regarding the photos are:

That the strongest and most valid testimony, that of the photographer, loses its definitely convincing character given the technical impossibility of proving if there was or not previous photographic montage.

That, finally the existence of personal testimonies and of a photographer, of some value given the circumstances involved, permit the admission that <u>there are indications of the existence of the UFO</u> [underlined in the original].[57]

In contrast to the careful and neutral style of the Brazilian Navy report, the U.S. Naval Attaché in Rio, the Office of Naval Intelligence (ONI), and Project Blue Book at the Air Technical Intelligence Center (ATIC) did not hesitate to label the Trindade Island UFO photos as a notorious hoax. The ONI Information Report from the Naval Attaché, while containing valuable data about the case and the position of

---

[57] Brazilian Navy report, ibid.

the Brazilian Navy, is written in a very negative-slanted style. It labels Barauna as a man with "a long history of photographic trick shots" and suggests that "the whole thing is a fake publicity stunt put on by a crooked photographer, and the Brazilian Navy fell for it." The coup de grace, however, is the final concluding remark by Captain Sunderland, USN:

> *It is the reporting officer's private opinion that a flying saucer would be unlikely at the very barren island of Trindade, as everyone knows Martians are extremely comfort loving creatures.*[58]

Dr. J. Allen Hynek, Northwestern University astronomer serving as consultant to Project Blue Book, observed appropriately that "such bias and flippancy have no place in scientific investigations."[59]

Likewise, Blue Book was quick to determine that "analysis of the Brazil picture by ATIC led to the conclusion that it was probably a hoax," although a "Record Card" admits that "this center [ATIC] has been unable to obtain copies of the photos."[60] This, despite the fact that Barauna's photos were widely available to the press and several UFO organizations both in Brazil and in the United States.

Olavo T. Fontes, the late pioneer Brazilian UFO investigator and medical doctor, compiled an extensive report on the case for the Aerial Phenomena Research Organization (APRO) in the United States, with transcripts of all the official statements and interviews published in the Brazilian

---

[58] Sunderland, M., Capt., USN, ONI Information Report re "Brazilian Navy—Flying Saucer Photographed from ALMIRANTE SALDANHA," March 11, 1958; reprinted in Hynek, Dr. J. Allen, *The Hynek UFO Report*, Dell, 1977.

[59] Hynek, ibid.

[60] ATIC documents in the Condon papers at the American Philosophical Society in Philadelphia.

press. Fontes disclosed additional sightings off Trindade Island, as well as other observations in the Atlantic Ocean from the Navy ships *Tridente* and *Triunfo*, and on the island of Fernando Noronha, located between Brazil and Africa, where "a U.S. guided missile and satellite tracking station" had just been set up. No official confirmation of the Noronha reports, however, was provided by Fontes.[61]

---

[61] Fontes, O.T., ibid.

1964: LANDING CASE
AT SOCORRO, NEW MEXICO

Until the experience of a small-town policeman in New Mexico, reports from persons claiming to have seen small beings in connection with UFOs on the ground (CE-III, or Close Encounters of the Third Kind) were looked upon with considerable disfavor within the UFO research community. After the landing near Socorro, New Mexico, confirmed by a second reputable witness, attitudes changed. The years following this event produced an unprecedented flow of reports of high credibility and strangeness.

At about 5:45 P.M. on Friday, April 24, 1964, Socorro policeman Lonnie Zamora was chasing a speeding car when his attention was drawn to a peculiar sight in the sky. "At this time I heard a roar and saw a flame in the sky to the southwest some distance away." Thinking it might be an explosion connected with a building known to contain explosives, he forgot about the car chase and sped off in the direction of the UFO.

The next time he saw it, it was on the ground, and from a distance it looked like a car that had overturned. As he drove closer, he saw that it resembled a large egg, sitting on one end and supported by slender legs. He stated:

> I saw two people in white coveralls very close to the
> object. One of these persons seemed to turn and look
> straight at my car and seemed startled—seemed to
> quickly jump somewhat. I don't recall noting any par-
> ticular shape or possibly any hats or headgear. These
> persons appeared normal in shape—but possibly they
> were small adults or small kids.

As he drove closer, a small hill blocked his view of the object, though at one point he heard a noise like that of a door closing. When he could again see the object, there was no one near it. He drove as close as the rough terrain would permit, stopped, parked his police cruiser, and got out, intending to walk toward the craft. At that point, "I heard about two or three loud 'thumps,' like someone possibly hammering or shutting a door or doors hard. These 'thumps' were possibly a second or less apart."

The white-suited individuals were not seen after he heard the thumps. As he started toward the object, it began to roar:

> *It started at a low frequency, but quickly the roar rose in frequency and in loudness . . . Flames were under the object . . . light blue and at bottom was a sort of orange color.*

Assuming it might be about to explode, Zamora quickly hid behind his cruiser for protection. The roaring then stopped and he looked up to see the craft hovering a few feet above the ground. "It was so quiet you could have heard a pin drop." The vehicle then moved away slowly, gathering speed as it headed toward the dynamite shack, which it cleared by a few feet.

At that time, Zamora was joined by a police sergeant who watched the craft fly away into the distance. Zamora and the sergeant then walked to where it had been parked, and noted charred and singed grass, underbrush, and imprints in the ground corresponding to where the vehicle had landed.[62]

Within hours, Zamora was interviewed by U.S. Army Captain Richard T. Holder, Up-Range Commander of the White Sands Missile Range, and by FBI Special Agent Arthur Byrnes, Jr., the latter requesting that the FBI's involvement be kept secret. Zamora described the object to them:

> *It was smooth—no windows or doors. As the roar started, it was still on or near the ground. There was*

---

[62] Written statement by Lonnie Zamora to Project Blue Book, 1964; reprinted in Steiger, Brad, ed., *Project Blue Book,* ibid.

*red lettering of some type. The insignia was about 2½
feet [75 cm.] high and about 2 feet [60 cm.] wide. It
was in the middle of the object. The object was . . .
aluminum-white.*

He then drew a sketch of the object with the red "insignia": half of a circle over an inverted V with a vertical line
inside and horizontal line below.[63]

A day or two later, Dr. J. Allen Hynek arrived to investigate the report for the Air Force's Project Blue Book. In
addition to questioning Zamora, Hynek measured and photographed the landing site. He located what appeared to be impressions in the ground made by the landing gear, as well as
several small footprints.

The case received rapid and extensive press coverage,
and the Air Force was under pressure to explain it as something less momentous than a landed spacecraft. Among the
explanations considered and rejected were a rancher's helicopter and an experimental NASA lunar lander.

In the end, Project Blue Book declared the report "unsolved," and Major Hector Quintanilla, the project's final
director, stated that there was no doubt that Lonnie Zamora
had seen an object that had left quite an impression on him:

*There is also no question about Zamora's reliability.
He is a serious police officer, a pillar of his church,
and a man well versed in recognizing airborne vehicles in his area. He is puzzled by what he saw, and
frankly, so are we. This is the best-documented case
on record, and still we have been unable, in spite of
thorough investigation, to find the vehicle or other
stimulus that scared Zamora to the point of panic.*[64]

---

[63] Ibid.

[64] Quintanilla, Hector, "The Investigation of UFO's," *Studies in Intelligence,* vol. 10, no. 4, fall 1966.

1967: PHYSIOLOGICAL CASE
AT FALCON LAKE, CANADA

The experience of Stephen Michalak in the Falcon Lake area in Manitoba, at noon on May 20, 1967, is a CE-II (Close Encounter of the Second Kind) on two counts: Physical traces were found on the area where the UFO reportedly landed, and the witness experienced a series of physiological effects apparently linked to his close encounter with a metallic-looking, disc-shaped object. Michalak is an industrial mechanic from Winnipeg who was doing some amateur prospecting in the area.

The case was investigated extensively by Canadian authorities, the Condon Commission, and several civilian UFO groups from the United States and Canada. The Royal Canadian Mounted Police (RCMP), the Department of National Defense (DND), the Royal Canadian Air Force (RCAF), and the Manitoba Department of Health were some of the agencies involved. Canadian officials reacted quickly after some radioactive traces were detected in soil samples from the landing area as well as on Michalak's garments. Many reports and documents on the case were eventually released by the Canadian government. One document provides a full summary of the case and investigation:

A Mr. Steven Michalak of Winnipeg, Manitoba reported that he had come into physical contact with a UFO during a prospecting trip in the Falcon Lake area, some 90 miles east of Winnipeg on 20 May 67. Mr. Michalak stated that he was examining a rock formation when two UFOs appeared before him. One of the UFOs remained airborne in the immediate area for a few moments, then flew off at great speed. The second UFO landed a few hundred feet away from his position. As he approached the UFO, a side door opened and voices were heard coming from within.

Mr. Michalak states he approached the object but was unable to see inside due to a bright yellow bluish light which blocked his vision. He endeavored to communicate with the personnel inside the object [in English, Russian, German, Italian, French, and Ukrainian], but without result. As he approached within a few feet of the object, the door closed. He heard a whining noise and the object commenced to rotate anti-clockwise and finally raised off the ground. He reached out with his left gloved hand and touched the object prior to its lifting off the ground; the glove burned immediately as he touched the object.

As the object left the ground, the exhaust gases burned his cap, outer and inner garments, and he sustained rather severe stomach and chest burns. As a result of these he was hospitalized for a number of days. The doctors who attended and interviewed Mr. Michalak were unable to obtain any information which could account for the burns to his body. The personal items of clothing which were alleged to have been burnt by the UFO, were subjected to an extensive analysis at the RCMP Crime laboratory. The analysts were unable to reach any conclusion as to what may have caused the burn damage.

Soil samples taken by Mr. Michalak from the immediate area occupied by the UFO were analyzed and found to be radioactive to a degree that the samples had to be safely disposed of. An examination of the alleged UFO landing area was made by a radiologist from the Department of Health and Welfare and a small area was found to be radioactive. The Radiologist was unable to provide an explanation as to what caused this area to become contaminated.

> *Both DND and RCMP investigation teams were*
> *unable to provide evidence which would dispute Mr.*
> *Michalak's story.[65]*

The RCAF investigation of Michalak, undertaken by Squadron Leader P. Bissky, was tough and highly skeptical. There were a few problems: Michalak failed to locate the landing site on two occasions when accompanied by the RCMP, but found it later with a friend. Much was made of this by physicist Roy Craig of the Condon Commission, who eventually dismissed the case.[66] However, Canadian researcher Chris Rutkowski makes a reasonable case of "disorientation in the wilderness" in discussing the details of the initial searches. Michalak had literally been taken from the hospital and flown in a helicopter by the RCAF to search for the spot. By the time of the third search, Michalak had partially recovered from his burns.[67]

S/L Bissky looked at the possibility of a hoax, searching for small details; for example, whether Michalak had handled "radium sources" at the cement company where he worked as a mechanic. Although S/L Bissky was trying to find holes in the story, he had to admit that "notwithstanding the evidence as it appears, the abdominal burns sustained by Mr. Michalak remain unexplainable as to the source of the burn."[68]

Michalak underwent several medical examinations in the

---

[65] "UFO Report—Falcon Lake, Man." Document in the RCMP case file; no author, agency, or department is identified.

[66] Gilmor, Daniel S., ibid.

[67] Rutkowski, Chris, "The Falcon Lake Incident," 3-part article published in *Flying Saucer Review*, July, August, and November 1981.

[68] Bissky, S/L P., "Report of an Investigation Into the Reported UFO Sighting by Mr. Stephen Michalak on May 20, 1967, in Falcon Lake Area," in the RCAF file.

course of the following months. The first took place on the evening of May 20 at Misericordia General Hospital in Winnipeg, where Michalak was taken by his son following his return from Falcon Lake on the day of the incident. The RCAF file includes a memorandum by a Deputy Base Surgeon who interviewed the physician who examined Michalak. The physician was not aware that the injuries were reportedly linked to a close encounter with a UFO, but had just been told that it was an accident. Surgeon D. J. Scott reported:

> At examination the physician found an area of first degree burns over the upper abdomen, covering an area of 7–8 inches [17–20 cm.] and consisting of several round and irregular shaped burns the size of a silver dollar or less. These were a dull red in color, the hair over the lower chest was singed as was the hair on the forehead with some questionable redness of the right cheek and temple.[69]

It is interesting to note that the geometrical burn marks on Michalak's chest and abdomen appear to conform to "a grid-like exhaust vent" observed by Michalak (see pages 81–82). According to Rutkowski's report:

> Unexpectedly, the craft shifted position, and he was now facing a grid-like exhaust vent which he had seen earlier to his left. A blast of hot air shot on to his chest, and set his shirt and undershirt on fire, and also caused severe pain. He tore off his burning garments, and threw them to the ground. He then looked up in time to see the craft depart and felt a rush of air as it ascended . . . He walked over to where he had left his things, and noticed that his compass was behaving erratically; after a few minutes, it became

---

[69] Scott, D. J., Deputy Base Surgeon, Memorandum to S/L P. Bissky, May 26, 1967; in the RCAF file.

> *still. He went back to the landing site, and imme-*
> *diately felt nauseous and a surge of pain from a*
> *headache.*[70]

Rutkowski summarized other physiological effects such as weight loss, "a drop of his blood lymphocyte count from 25 to 16 percent," swelling of his body, and other ailments. He described as well the circumstances surrounding a series of physical and psychiatric tests undertaken by Michalak at the Mayo Clinic in Rochester, Minnesota, in 1968, at his own expense. Since Michalak was found in general good health, normal medical explanations such as neurodermatitis and hyperventilation were hypothesized. The psychiatrists determined that, despite the stress caused by all the publicity generated by his UFO experience, "there was no other evidence of delusions, hallucinations, or other emotional disorders."[71]

One of the weaknesses of the case is that Michalak was the only witness. No one corroborated his crucial testimony of the landing or overflight of a disc-shaped object. Professor Craig chose to dismiss the whole incident with curious reasoning in his final "Conclusion of 'Case 22' " for the Condon Report:

> *If Mr. A's [Michalak] reported experience were physi-*
> *cally real, it would show the existence of alien flying*
> *vehicles in our environment. Attempts to establish the*
> *reality of the event revealed many inconsistencies*
> *and incongruities in the case, a number of which are*
> *described in this report. Developments subsequent to*
> *the field investigation have not altered the initial con-*
> *clusion that this case does not offer probative infor-*
> *mation regarding unconventional craft.*[72]

---

[70] Rutkowski, C., ibid.

[71] Ibid.

[72] Craig, Roy, "Case 22 North Central Spring 67," in Gilmor, D.S., ibid.

Yet, a careful review of all the physical and medical evidence collected by the RCMP and others could easily lead one to the opposite conclusion. Moreover, some of the physiological effects reported in the Falcon Lake incident are not isolated events in the UFO literature. Aerospace engineer John Schuessler has been documenting UFO medical cases for many years, compiling a Catalog of Medical Injury Cases. The 1995 version of the catalog contains approximately 400 cases.[73] Although this particular field requires further research, it is one area where at least "partial proof" can be offered.

---

[73] Schuessler, John F., "Developing a Catalog of UFO-Related Human Physiological Effects," *MUFON 1995 International UFO Symposium Proceedings.*

1975: STRATEGIC AIR COMMAND BASES UFO ALERT

Visual sightings and radar tracking of UFOs in the vicinity of military installations have been reported since the beginning of the modern era. The number and nature of most of these events has been kept from the public by military security, but on occasion information has been released, although its significance is usually played down.

From late October through the middle of November 1975, high-security bases along the U.S.–Canada border were the scene of intrusions of what were euphemistically called "mystery helicopters," despite their unhelicopterlike appearance and behavior.

The most complete recounting of the events from anyone in the U.S. or Canadian governments is from the Commander-in-Charge of the North American Aerospace Defense Command (NORAD), on November 11, 1975:

> *Part I. Since 28 Oct. '75, numerous reports of suspicious objects have been received at the NORAD CU. Reliable military personnel at Loring AFB, Maine, Wurtsmith AFB, Michigan, Malmstrom AFB, Montana, Minot AFB, North Dakota, and Canadian Forces Station Falconbridge, Ontario, Canada, have visually sighted suspicious objects.*

> *Part II. Objects at Loring and Wurtsmith were characterized to be helicopters. Missile Site Personnel, Security Alert Teams, and Air Defense Personnel at Malmstrom AFB, Montana report an object which sounded like a jet aircraft. FAA advised there were no jet aircraft in the vicinity. Malmstrom search and height finder radars carried the object between 9,500 ft. [2,900 m.] and 15,600 ft. [4,850 m.] at a*

speed of 7 knots [9 mph or 14 km./hr.]. There was in-termittent radar contact with the object from 080753Z through 0900Z November 1975. F-106's scrambled from Malmstrom could not make contact due to dark-ness and low altitude. Site personnel reported the ob-ject as low as 200 ft. [60 m.] and said that as the interceptors approached, the lights went out. After the interceptors had passed, the lights came on again, one hour after the F-106's returned to base. Missile site personnel reported the object increased to a high speed, raised [sic] in altitude and could not be dis-cerned from the stars.

Part III. Minot AFB on 10 Nov. reported that the site was buzzed by a bright object the size of a car at an altitude of 1,000 to 2,000 ft. [300 to 600 m.]. There was no noise emitted by the vehicle.

Part IV. This morning, 11 Nov. '75 CFS Falconbridge reported search and height finding radar paints on an object 25 to 30 nautical miles [30 to 35 statute miles or 48 to 56 km.] south of the site ranging in altitude from 26,000 ft. to 72,000 ft. [6,500 m. to 18,000 m.]. The site commander and other personnel say the object appeared as a bright star but much closer. With binocu-lars, the object appeared as a 100 ft. diameter sphere and appeared to have craters around the outside.

Part V. Be assured that this command is doing every-thing possible to identify and provide solid factual in-formation on these sightings. I have also expressed my concern to SAFOI [Secretary of the Air Force, Of-fice of Information] that we come up soonest with a proposed answer to queries from the press to prevent over reaction by the public to reports by the media that may be blown out of proportion. To date, efforts

*by Air (National) Guard helicopters, SAC helicopters
and NORAD F-106s have failed to produce positive
ID [identification].[74]*

The most important fact omitted from the NORAD report was that many of the sightings were of objects over or
near areas used for the storage of nuclear weapons. This fact
is clearly established by other declassified documents that
mention that, "in the interest of nuclear weapons security the
action addressees will assume Security Option III during
hours of darkness until further notice." A SAC message on
the subject of "Defense Against Helicopter Assault" captures
the alert mood:

*Several recent sightings of unidentified aircraft/heli-
copters flying/hovering over Priority A restricted
areas during the hours of darkness have prompted the
implementation of security Option 3 at our northern
tier bases. Since 27 Oct. 75, sightings have occurred
at Loring AFB, Wurtsmith AFB, and most recently, at
Malmstrom AFB. All attempts to identify these aircraft have met with negative results.[75]*

With the exception of the object reported from CFS
Falconbridge, only limited descriptions were given of the
appearance of the UFOs. From a variety of military sources
come these descriptions:

*October 28, Loring AFB, Maine. Unknown craft with
a white flashing light and an amber or orange light.
Red and orange object, about four car-lengths long.*

---

[74] Official U.S. Air Force report, NORAD, November 11, 1975.

[75] CINCSAC Offutt AFB message, "Subject: Defense Against Helicopter Assault," November 10, 1975.

Moving in jerky motions, stopped and hovered. The object looked like all the colors were blended together; the object was solid.

October 30, Wurtsmith AFB, Michigan. One light pointing downward, and two red lights near the rear. Hovered and moved up and down in an erratic manner. A KC-135 aerial tanker crew established visual and radar contact with UFO: "Each time we attempted to close on the object, it would speed away from us. Finally, we turned back in the direction of the UFO and it really took off . . . I know this might sound crazy, but I would estimate that the UFO sped away from us doing approximately 1,000 knots [1,150 mph or 1,800 km./hr.]."

November 7, Malmstrom AFB, Montana. A Sabotage Alert Team described seeing a brightly glowing orange, football-field-sized disc that illuminated the Minuteman ICBM missile site. As F-106 jet interceptors approached, the UFO took off straight up, NORAD radar tracking it to an altitude of 200,000 feet [38 miles or 60 km.]. An object . . . emitted a light which illuminated the site driveway. The orange-gold object overhead also has small lights on it.

November 8, Malmstrom AFB. Radar showed up to 7 objects at 9,500 to 15,000 feet [2,900 m. to 4,700 m.]. Ground witnesses reported lights and the sound of jet engines, but radar showed objects flying at only 7 kts. [8 mph or 13 km./hr.].

November 10, Minot AFB, North Dakota. A bright, noiseless object about the size of a car buzzed the base at 1,000–2,000 feet [300–600 m.].[76]

---

[76] Extracts: 24 NORAD Region Senior Director Log November 1975.

1976: MULTIPLE WITNESS CASE IN THE CANARY ISLANDS

A large, unidentified luminous phenomenon was observed throughout the Spanish Canary Islands on the night of June 22, 1976. Newspaper headlines proclaimed the following day that "thousands of people" had seen a "spectacular luminous phenomenon" that "lasted twenty minutes and was observed from Tenerife, La Palma and La Gomera."[77] The most sensational aspect was the experience of a medical doctor and his taxi driver, who reported a transparent sphere with two tall entities inside.

On June 25, 1976, the Commanding General of the Canaries' Air Zone named an "Investigative Adjutant" to investigate the case. Copies of some of the depositions, though technically confidential, were given by a Spanish Air Force General to journalist J. J. Benitez in October 1976, who subsequently published them in his book *UFOs: Official Documents of the Spanish Government.*[78] The complete Air Force file on the case, comprising over one hundred pages of questionnaires, evaluations, appendices, illustrations, etc., was officially declassified in June 1994, as part of the ongoing public release of the Spanish Air Force UFO files, which began in 1992.

The Air Force file contains depositions with fourteen witnesses. In a previously established methodology, it divides observers into four categories according to their reliability: from high credibility rating (pilots, aeronautical engineers, astronomers) all the way to a very unreliable rating (those who were illiterate, mentally impaired, or under the influence of

---

[77] Benítez, J. J., *OVNIS: Documentos Oficiales del Gobierno Español,* Barcelona, Plaza & Janes, 1977.

[78] Ibid.

alcohol or drugs). Likewise, each observation is also divided into four levels, according to the number and quality of additional witnesses, radar evidence, etc. The dossier also included a color photograph of the phenomenon, taken by a tourist and obtained with the assistance of the Civil Guard; according to the photo lab where the film was developed, "no trickery or modification of any kind was added."

The Investigative Adjutant reconstructed the sequence of events in his final report. The first observation was made at 21:27 hrs. on June 22, 1976, by the entire crew of the corvette *Atrevida* of the Spanish Navy, which was located 3 nm (3.5 statute miles or 5.5 km.) off Punta Lantailla on the coast of Fuerteventura Island. The ship's captain provided a detailed description of the event:

At 21:27 (Z) hrs. on 22 June, we saw an intense yellowish-bluish light moving out from the shore towards our position. At first we thought it was an aircraft with its landing lights on. Then, when the light had attained a certain elevation (15–18 degrees), it became stationary. The original light went out and a luminous beam from it began to rotate. It remained like this for approximately two minutes. Then an intense great halo of yellowish and bluish light developed, and remained in the same position for 40 minutes, even though the original phenomenon was no longer visible.

Two minutes after the great halo, the light split into two parts, the smaller part being beneath, in the center of the luminous halo, where a blue cloud appeared and the part from which the bluish nucleus had come, vanished. The upper part began to climb in a spiral, rapid and irregular, and finally vanished. None of these movements affected the initial circular halo in any way, which remained just the same the whole time, its glow lighting up parts of the land and the

ocean, from which we could deduce that the phenome-
non was not very far away from us, but was close.[79]

The file also includes the deposition of a lieutenant, the
Atrevida's first officer, and indicates "the phenomenon was
initially observed by the entire crew" of the Navy warship.
The report adds that no echo was detected on the ship's surface radar. Three minutes later, at 21:30 hrs., a very similar
phenomenon was observed by many people in Grand Canary
Island. The majority of the witnesses interviewed by the
Air Force were from the villages of Galdar, Las Rosas, and
Agaete. They were from different professions: medical doctor, schoolteacher, farmer, sergeant, two taxi drivers, police
guard, and laborers. Newspapers and UFO investigators located additional witnesses in the islands.

The Investigative Adjutant determined that there was no
aerial traffic or military exercises at the time that could account for the phenomenon. The observation itself was divided into two categories: the large luminous halo in the
sky, seen by many people; and the smaller luminous globe
with two figures inside, observed by a doctor, a taxi driver,
and one woman. The Adjutant had no problem accepting the
reality of the first event. Noting that it was vouched for
officially by the crew of the *Atrevida*, he added:

Then, numerous witnesses belonging to different po-
sitions and cultural strata, saw it with similar char-
acteristics in the Grand Canary island. Therefore, the
fact that a very strange and peculiar aerial phenome-
non occurred on the night of 22 June is a true and
proven fact, as incredible as its behavior and condi-
tions may seem.

---

[79] Deposition No. B-07 of the Captain of Corvette in the Spanish Air
Force file; English translation by Gordon Creighton, *Flying Saucer
Review,* vol. 23, no. 3, 1977.

The Adjutant considered four possible explanations—aircraft, missile test, aurora, and meteor fall—only to reject each hypothesis, one at a time. The report also considered and rejected other explanations such as weather balloons and meteorological phenomena, admitting that "its nature is totally unknown."[80]

The Investigative Adjutant, however, had more problems accepting the reality of the CE-III (Close Encounter of the Third Kind) described by some of the witnesses. Not because he questioned their veracity or suspected them of hoaxing, but simply because of the nature of the report. The CE-III's main witness was a physician from the town of Guía, Dr. Francisco Padrón León. His deposition is the longest in the file. The Air Force had also checked his background and psychological condition. Dr. Padrón explained that he had been called to attend a patient and was riding in a taxi to see her in the town of Las Rosas:

> We were talking about hunting . . . as we entered the
> last part of the road, the car lights pointed at a
> slightly luminous sphere that was stationary and very
> close to the ground, although I can't say for sure if it
> was touching it; it was made of a totally transparent
> and crystalline-like material, since it was possible to
> see through it the stars in the sky; it had an electric
> blue color but tenuous, without dazzling; it had a ra-
> dius of about 30 m. [100 ft.], and in the lower third of
> the sphere you could see a platform of aluminum-like
> color as if made of metal, and three large consoles. At
> each side of the center there were two huge figures of
> 2.50 to 3 m. [8.5 to 10 ft.] tall, but no taller than 3 m.
> [10 ft.], dressed entirely in red and facing each other
> in such a way that I always saw their profile.

---

[80] Adjutant's Report, Las Palmas Aerial Sector, July 16, 1976.

They were humanoid in shape, with the head proportionate to the thorax and wearing some kind of headgear. Dr. Padrón asked the taxi driver if he was seeing the same thing, and he exclaimed, "My God! What is that?" As the car reached the patient's house, the doctor noted:

Then I observed that some kind of bluish smoke was coming out from a semi-transparent central tube in the sphere, covering the periphery of the sphere's interior without leaking outside at any moment. Then the sphere began to grow and grow until it became huge like a 20-story house, but the platform and the crew remained the same size; it rose slowly and majestically and it seems I heard a very tenuous whistling.

Dr. Padrón entered the house and alerted the residents, who went outside and saw:

The sphere, now high, was moving slowly toward Tenerife; suddenly it reached enormous speed like none I ever saw in an airplane; the sphere dissolved into a bluish spindle-shape with red underneath; a brilliant white halo was formed close to the object, which bit by bit was forming another very brilliant blue halo. It disappeared in the direction of Tenerife.[81]

Dr. Padrón's testimony was confirmed by the taxi driver, who also saw "a craft that looked as if it was made of transparent crystal," about 25 m. (85 ft.) high and 20 m. (65 ft.) wide, with "two persons dressed in brilliant red inside." In addition, there was a third witness, an illiterate woman who

---

[81] Deposition No. A-01 by Dr. Francisco Padrón León in the Spanish Air Force file.

was a relative of the doctor's patient. She was watching TV when the screen went blank and the dogs began to bark. She ran to the window in time to:

> . . . see the doctor's car and just above it the great blue ball . . . It was like a perfectly round globe, but very big, transparent, the stars could be seen through it. She saw two man-like figures inside, but she is not completely sure as she panicked, closed the windows and doors of the house and began to pray.[82]

Because of the strange nature of the CE-III, the Investigative Adjutant had reservations accepting it. He remarked:

> We should forcefully consider the VERY PROBABLE [sic] circumstance that both witnesses, facing the presence of an unusual phenomenon in the sky, narrated what their "minds" made them see, mutually influencing each other. This Investigative Adjutant doesn't have the slightest doubt about their seriousness and sincerity. They told what they unquestionably "believed" to have seen.[83]

The Adjutant's final conclusion, however, was that what the crew of the corvette *Atrevida* and many other witnesses observed in the sky on the night of June 22 was indeed an "Unidentified Aerial Phenomenon." It is important to note that this incident was neither the first nor the last UFO report investigated officially in the Canary Islands. On November 19, 1976, the Commanding General of the Canaries' Air Zone, Carlos Dols de Espejo, and his aides observed firsthand another large halo while flying on an Air Force T-12

---

[82] Deposition No. A-02 by taxi driver and No. B-05 by woman in Galdar, in the Spanish Air Force file.

[83] Adjutant's Report, ibid.

transport plane. The crew of a Spanish Navy training ship and the personnel at the Gando Air Base also reported the phenomenon. The Investigative Adjutant in that case concluded his report:

> *If we study as a whole the three reports issued up to the present (1/75, 1/76, and 2/76), we should have to think seriously of the necessity of considering the possibility of accepting the hypothesis that a craft of unknown origin, propelled by an equally unknown energy, is moving freely over the skies in the Canaries.*[84]

[84] Benítez, J. J., "Informe 02/76 de las Fuerzas Aéreas españolas," *Mundo Desconocido,* Barcelona, September 1979; quoted in Huneeus, Antonio, "Top Spanish General sees UFO," *The News World,* New York, December 5, 1981.

1976: UFO DOGFIGHT OVER TEHRAN

One of the best-documented UFO incidents took place over the skies of Tehran, Iran, on the night of September 18–19, 1976, when a UFO was observed flying over the capital's restricted airspace. Two U.S.–made F-4 Phantom II jet fighters of the Imperial Iranian Air Force were scrambled, but as the pilots closed in on the target, their communications and weapons systems were suddenly jammed. The incident was confirmed by high-ranking officers of the Iranian Air Force and later documented by several agencies of the U.S. military.

Lt. General Azarbarzin, Deputy Commander-in-Chief of Operations of the Imperial Iranian Air Force, confirmed in a 1977 interview the strong electromagnetic effect experienced by the two F-4s:

> *That is true. They both were scrambled and they locked on the target but they received a very strong jamming. And then they lost almost every avionics system they had on the airplane . . . The jets couldn't fire their missiles because they had very strong jamming . . . this technology it [UFO] was using for jamming was something we haven't had before and we don't have it. It doesn't exist because it was a very wide band and could jam different bands, different frequencies at the same time. It's very unusual.*[85]

General Nader Yousefi, Base Operations Commander and the number-three man in the Imperial Iranian Air Force, authorized the scramble mission and was also an eyewitness, as described in a TV interview:

---

[85] Cathcart, John, transcript of interview with Lt. Gen. Azarbarzin, January 4, 1977, filed with the Fund for UFO Research.

*I put down my phone [with the Control Tower] and I
ran to my balcony to see if I can see that object. I saw
a big star among the other stars, which it was at least
twice as large as the normal stars . . . It was around
12 miles [19 km.], we lost communication and I heard
nothing from the pilots, so I was so scared what's
going to happen and what happened to the pilots. I
asked from the tower controller to tell them to continue their mission and see if they can get more information from the flying object . . . and it [UFO] was
coming toward them, they try to shoot them down, when
they squeezed the trigger it didn't work and the trigger was inoperative, they couldn't shoot the missiles.*[86]

The sequence of events can be reconstructed both from
Iranian sources and from declassified U.S. intelligence
documents:

Between 10:30 and 11:15 P.M. on September 18, several
calls were received by the Control Tower at Mehrabad Airport, reporting an unknown object hovering 1,000 feet (300 m.)
above the ground in the northern section of Tehran. The
night-shift supervisor, Hossain Perouzi, initially didn't pay
too much attention. The radar system was turned off since it
was under repair. After he received the fourth telephone call,
at 11:15 P.M., Perouzi went to the terrace next to the tower
and observed the UFO with binoculars:

*Suddenly I saw it. It was rectangular in shape, proba-
bly seven to eight meters [24 to 27 ft.] long and about
two meters [7 ft.] wide. From later observations I
made, I would say it was probably cylindrical. The
two ends were pulsating with a whitish blue color.
Around the mid-section of the cylinder there was this*

---

[86] International Noor Productions, Sherman Oaks, California, video-
taped interview with Gen. Yousefi, shown in the TV program *Sightings*, 1994.

*small red light that kept going around in a circle ... I
was amazed. I didn't know what to think. There definitely was a very strange object there in the sky right
over Tehran.*[87]

At 12:30 A.M. on September 19, Perouzi called the Air
Force Command post to report the UFO. The Base Commander, in turn, called General Yousefi, who authorized the
scramble of an F-4 Phantom jet from Shahrokhi AFB to investigate. The first scramble is summarized in a "Memorandum for Record" from the U.S. Defense Attaché Office
(DAO) in Tehran:

> *The F-4 took off at 01:30 ... and proceeded to a point
> about 40 nm [45 statute miles or 75 km.] North of
> Tehran. Due to its brilliance the object was easily
> visible from 70 miles [110 km.] away. As the F-4
> approached a range of 25 nm [29 statute miles or
> 46 km.], he lost all instrumentation and communica-
> tions (UHF and Intercom). He broke off the intercept
> and headed back to Shahrokhi. When the F-4 turned
> away from the object and apparently was no longer a
> threat to it, the aircraft regained all instrumentation
> and communications.*[88]

At 01:40 hours, Gen. Yousefi authorized a second F-4
scramble piloted by Lt. Jafari, who quickly established radar
contact with the UFO. The DAO Memorandum describes
the events of the second scramble:

---

[87] Petrozian, transcript of interview with Hosain Perouzi, December
22, 1976, filed with the Fund for UFO Research.

[88] Mody, Lt. Colonel Olin, USAF, Memorandum for Record, "Subject: UFO Sighting," undated; the text appears virtually identical to
an "unclassified" message from the Defense Attaché Office in
Tehran, September 23, 1976.

> *The size of the radar return was comparable to that of a [Boeing] 707 tanker. The visual size of the object was difficult to discern because of its intense brilliance. The light that it gave off was that of flashing strobe lights arranged in a rectangular pattern of alternating blue, green, red and orange, in color. The sequence of the lights was so fast that all the colors could be seen at once.*
>
> *The object and the pursuing F-4 continued a course to the south of Tehran when another brightly lighted object, estimated to be ½ to ⅓ the apparent size of the moon, came out of the original object. The second object headed straight toward the F-4 at a very fast rate. The pilot attempted to fire an AIM-9 missile at the object, but at that instant his weapons control panel went off and he lost all communications (UHF and Interphone). At this point, the pilot initiated a turn and negative G dive to get away.[89]*

The aircraft electric system went back to normal once the F-4 reached a certain distance from the UFO. The small object returned to the primary object, but a second one emerged and flew toward the ground. Gen. Yousefi observed the landing from the balcony of his Tehran residence:

> *He went down and landed on the ground and now it is a communication between the mothership and that small flying object, and it shows the lights between those two is connected.[90]*

Still more strange events were reported that night. A UFO seemed to follow the F-4 as it approached the runway,

---

[89] Ibid.

[90] International Noor Productions, ibid.

and a civil airliner experienced communications failure but did not see anything. The DAO Memorandum describes the investigation early that morning:

> *During daylight, the F-4 crew was taken out to the area in a helicopter where the object apparently had landed. Nothing was noticed at the spot where they thought the object landed (a dry lake bed), but as they circled off to the West of the area they picked up a very noticeable beeper signal. At that point, where the return was the loudest was a small house with a garden. They landed and asked the people within if they had noticed anything strange last night. The people talked about a loud noise and a very bright light like lightning.[91]*

The trail ends there. Although the Attaché Office added that the area had been checked for possible radiation and that "more information will be forwarded when it becomes available," the details surrounding the beeper signal and the ground witnesses have not been released. A report published in a 1978 classified U.S. military journal, *MIJI Quarterly,* basically repeats the facts contained in the original DAO message, although its author begins the article with this interesting remark:

> *Sometime in his career, each pilot can expect to encounter strange, unusual happenings which will never be adequately or entirely explained by logic or subsequent investigation. The following article recounts such an episode as reported by two F-4 Phantom crews of the Imperial Iranian Air Force during late 1976. No additional information or ex-*

---

[91] Mody, ibid.

planation of the strange events has been forthcoming; the story will be filed away and probably forgotten, but it makes interesting, and possibly disturbing, reading.[92]

Recent taped testimonies by Iranian Air Force Generals Nader Yousefi and Mahmoud Sabahat, now retired and living in exile in the United States, reveal that General John Secord, then chief of the USAF mission in Iran, attended a high-level briefing with Iranian authorities and the pilots and air traffic controllers involved in the incident. A Defense Intelligence Agency (DIA) "Evaluation" summarized the salient features of the Iranian incident:

An outstanding report. This case is a classic which meets all the criteria necessary for a valid study of the UFO phenomenon:

a) The object was seen by multiple witnesses from different locations (i.e., Shemiram, Mehrabad and the dry lake bed) and viewpoints (both airborne and from the ground).

b) The credibility of many of the witnesses was high (an Air Force General, qualified aircrews, and experienced radar operators).

c) Visual sightings were confirmed by radar.

d) Similar electromagnetic effects (EME) were reported by three separate aircraft.

---

[92] Shields, Captain Henry, USAFE, "Now You See It, Now You Don't," United States Air Force Security Service, *MIJI Quarterly*, October 1978.

e) There were physiological effects on some crew members (i.e., loss of night vision due to the brightness of the object).

f) An inordinate amount of maneuverability was displayed by the UFOs.[93]

---

[93] Defense Information Report Evaluation, DIA, October 12, 1976; reprinted in *The UFO Cover-up*, by Lawrence Fawcett and Barry Greenwood, Simon & Schuster, 1992.

1980: UFO INCIDENTS AT RENDLESHAM FOREST, ENGLAND

Several UFO incidents, including multiple-witness sightings by military personnel and ground traces with above-normal radioactive readings, were reported in late December 1980 at the Rendlesham Forest in Suffolk, England. The site was near two then-important NATO bases leased to the U.S. Air Force: RAF Bentwaters and RAF Woodbridge. Although details and dates reported by various investigators in the 1980s are somewhat confusing, there is an official record of the case in a memorandum to the British Ministry of Defence (MOD), signed by USAF Lt. Col. Charles I. Halt, Bentwaters Deputy Base Commander:

SUBJECT: Unexplained Lights
TO: RAF/CC

1. Early in the morning of 27 December, 1980 (approximately 0300L, or 3 A.M. local time), two USAF security police patrolmen saw unusual lights outside the back gate at RAF Woodbridge. Thinking an aircraft might have crashed or been forced down, they called for permission to go outside the gate to investigate. The on-duty flight chief responded and allowed three patrolmen to proceed on foot. The individuals reported seeing a strange glowing object in the forest. The object was described as being metallic in appearance and triangular in shape, approximately two to three meters [7 to 10 ft.] across the base and approximately two meters [6.5 ft.] high. It illuminated the entire forest with a white light. The object itself had a pulsing red light on top and a bank of blue lights underneath. The object was hovering or on legs. As the patrolmen approached the object, it maneuvered through the trees and disappeared. At this time, the animals on a nearby farm went into a frenzy. The object was briefly sighted approximately an hour later near the back gate.[94]

In addition to Col. Halt's summary, testimony was provided by the USAF patrolmen involved in the case. Law enforcement airman John Burroughs wrote an official deposition of his experience after spotting some lights while on patrol near Woodbridge's East Gate:

We stopped the truck where the road stopped and went on foot. We crossed a small open field that lead [sic] into the trees where the lights were coming from and as we were coming into the trees there were strange noises, like a woman was screaming, also the woods lit up and you could hear the farm animals making a lot of noise and there was a lot of movement in the woods. All three of us hit the ground and whatever it was started moving back towards the open field and after a minute or two we got up and moved into the trees and the lights moved out into the open field.[95]

Burroughs drew a sketch of the object in his official statement In a 1990 interview, Burroughs described the object as:

A bank of lights, differently colored lights that threw off an image of like-a-craft. I never saw anything metallic or anything hard.

---

[94] Halt, Lt. Col. Charles I., USAF, Memorandum to MOD, "SUBJECT: Unexplained Lights," January 13, 1980.

[95] Huneeus, Antonio, "The Testimony of John Burroughs," *Fate,* September 1993.

Yet, the most interesting part of his testimony is not the presence of the lights, but rather his sensation of an altered state of consciousness:

> *Everything seemed like it was different when we were in that clearing. The sky didn't seem the same . . . it was like a weird feeling, like everything seemed slower than you were actually doing, and all of a sudden when the object was gone, everything was like normal again.*[96]

The testimonies of Burroughs and the other members of the USAF security patrol were confirmed the following day by the finding of ground traces with radioactive readings in the forest. Col. Halt summarized the events in his memorandum to the MOD:

> *2. The next day, three depressions 1½ feet [.5 m.] deep and 7 feet [2 m.] in diameter were found where the object had been sighted on the ground. The following night (29 December, 1980) the area was checked for radiation. Beta/gamma readings of 0.1 milliroentgen were recorded with peak readings in the three depressions and near the center of the triangle formed by the depressions. A nearby tree had moderate (.05–.07) readings on the side of the tree toward the depressions.*[97]

Col. Halt, moreover, became directly involved in the UFO incidents when he led a second patrol into the forest two nights later. He made an audiotape recording describing live the puzzling events of that night. While the tape runs for

---

[96] Ibid.

[97] Halt, C. I., ibid.

about twenty minutes, it covers a span of over three hours, so there are obviously cuts in between. The tape describes their efforts to carry on the radiation readings quoted above and, as the night goes on, the voices become increasingly excited as strange lights appear in the forest:

> *OK, we're looking at the thing, we're probably about two or three hundred yards. It looks like an eye wink-ing at you. It's still moving from side to side and when you put the starscope [a night-vision device] on it, it's like this thing has a hollow center, a dark center. It's a bit like a pupil of an eye looking at you, winking, and the flash is so bright through the starscope that it al-most burns your eye.[98]*

Col. Halt summarized these events in the third part of his memo to the MOD:

> *3. Later in the night, a red sun-like light was seen through the trees. It moved about and pulsed. At one point, it appeared to throw off glowing particles and then broke into five separate white objects and then disappeared. Immediately thereafter, three star-like objects were noticed in the sky, two objects to the north and one to the south, all of which were about 10 degrees off the horizon. The objects moved rapidly in sharp angular movements and displayed red, green and blue lights. The objects to the north appeared to be elliptical through an 8–12 power lens. They then turned to full circles. The objects to the north re-mained in the sky for an hour or more. The object to the south was visible for two or three hours and beamed down a stream of light from time to time. Nu-*

---

[98] Transcript of Col. Halt's audio recording published in Good, T., *Above Top Secret,* Quill, William Morrow, 1988.

*merous individuals, including the undersigned, wit-
nessed the activities in paragraphs 2 & 3.*[99]

Charles Halt discussed the case again after retiring from
the USAF with the rank of full colonel. He told the TV program *Unsolved Mysteries* in 1991:

*I was very skeptical. I found what allegedly had taken
place hard to believe, and I was really going to de-
bunk it quite frankly; and as events unfolded I became more and more concerned that there maybe is
something to this . . . I kept telling myself that there
had to be some type of explanation for it, but I certainly couldn't find one and even to this day I can't explain what happened.*

Col. Halt alluded to the military implications of the event
when describing beams from the object pointing to the
weapons-storage area at Woodbridge:

*We could very clearly see it . . . I noticed other beams
of light coming down from the same object falling on
different places on the base. My boss was standing in
his front yard in Woodbridge and he could see the
beams of light falling down, and the people in the
weapons storage area and other places on the base
also reported the lights.*[100]

Many accounts and commentaries have been published
on the Rendlesham Forest and Bentwaters incidents. Some,

---

[99] Halt, C. I., ibid.

[100] Interview with Col. (Ret.) Halt, *Unsolved Mysteries,* "U.S. military officers discuss a 1980 sighting of an unidentified flying craft
near a U.S. air base in England," originally broadcast on NBC-TV
on September 18, 1991.

like the appearance of ghostlike entities, are still enveloped by controversy. The Rendlesham events have been mentioned in both the House of Commons and the House of Lords, and looked into by Nebraska Senator James Exon.[101]

A reexamination of the Rendlesham Forest UFO incidents was undertaken recently by Nick Pope during his three-year tour as head of the MOD Secretariat Air Staff (AS2) office, which inherited the UFO reporting function from DS8. Mr. Pope checked with radiation experts as to the significance of the 0.1 milliroentgen of beta/gamma readings taken by Col. Halt's patrol:

> I went to an organization called the Defense Radiological Protection Service, which is a unit attached to the Institute of Naval Medicine near Gosport, Hampshire, and they told me that the levels of radiation reported by Col. Halt in that memo were ten times what they should be in that area compared to their background samples.[102]

British author and researcher Ralph Noyes was for four years the head of Defense Secretariat 8 (DS8), retiring in 1977 with the rank of Under Secretary of State. He wrote regarding this case:

> Our worried skeptical colleagues have already had to advance an extraordinary hotch-potch of explanations: space debris, a bright meteor, a police car, drink and drugs, a lighthouse, other lights on the coast, dear old Sirius.

---

[101] See Butler, Street & Randles, *Sky Crash*, Neville Spearman, 1984; Randles, J., *From Out of the Blue*, Global Communications, 1991; "The Bentwaters Incident," articles by Jenny Randles, Ray Boeche, and Antonio Huneeus, *Fate*, September 1993.

[102] Pope, Nick, lecture at the New Hampshire MUFON Conference, Portsmouth, September 10, 1995.

> *Occam, you will remember, urged us to cut away unnecessary complications in our attempts to explain phenomena and to look for the simplest explanation. The simplest explanation of Halt's memorandum is that he was reporting—as precisely as wondrous events permit—what he and "numerous individuals" encountered on December 29/30, together with such facts as he had been able to ascertain from his subordinates about the occurrences of December 26/27.*[103]

---

[103] Noyes, Ralph, "UFO lands in Suffolk—and that's Official," chapter in Timothy Good's anthology, *The UFO Report 1990,* Sidgewick & Jackson, 1989.

1981: PHYSICAL TRACE CASE IN TRANS-EN-PROVENCE, FRANCE

On the afternoon of January 8, 1981, a strange craft landed on a farm near the village of Trans-en-Provence in the Var region in southeastern France. Physical traces left on the ground were collected by the gendarmerie within twenty-four hours and later analyzed in several French government laboratories. Extensive evidence of anomalous activity was detected.

The case was investigated by the Groupe d'Etudes des Phénomènes Aérospatiaux Non-identifiés (GEPAN), or Unidentified Aerospace Phenomena Study Group, established in 1977 within the National Center for Space Studies (CNES) in Toulouse, the French counterpart of NASA. (The functions of GEPAN were reorganized in 1988 into the Service d'Expertise des Phénomènes de Rentrées Atmosphériques, or SEPRA). The primary investigator was Jean-Jacques Velasco, the head of SEPRA.

The witness was the farmer Renato Nicolai, age 55, on whose property the UFO landed and then took off almost immediately. Thinking that it was a military experimental device, Nicolai notified the local gendarmes on the following day. The gendarmes interviewed Nicolai and collected soil and plant samples from the landing site within twenty-four hours of the occurrence, notifying GEPAN on January 12 as part of a cooperation agreement for UFO investigation between the two agencies. Further collection of samples and measurements of the site were undertaken by the GEPAN team, and the samples were thoroughly analyzed by several government laboratories.

The first detailed report on the case was published by GEPAN in 1983. Nicolai's testimony to the police was simple and straightforward:

My attention was drawn to a small noise, a kind of lit-tle whistling. I turned around and I saw, in the air, a ship which was just about the height of a pine tree at the edge of my property. This ship was not turning but was descending toward the ground. I only heard a slight whistling. I saw no flames, neither underneath or around the ship.

While the ship was continuing to descend, I went closer to it, heading toward a little cabin. I was able to see very well above the roof. From there I saw the ship standing on the ground.

At that moment, the ship began to emit another whistling, a constant, consistent whistling. Then it took off and once it was at the height of the trees, it took off rapidly . . . toward the northeast. As the ship began to lift off, I saw beneath it four openings from which neither smoke nor flames were emitting. The ship picked up a little dust when it left the ground.

I was at that time about 30 meters [100 feet] from the landing site. I thereafter walked towards the spot and I noticed a circle about two meters [7 feet] in di-ameter. At certain spots on the curve of the circle, there were tracks (or traces).

The ship was in the form of two saucers upside down, one against the other. It must have been about 1.5 meters [5 feet] high. It was the color of lead. The ship had a border or type of brace around its circum-ference. Underneath the brace, as it took off, I saw two kinds of round pieces which could have been landing gear or feet. There were also two circles which looked like trap doors. The two feet, or landing gear, extended about 20 centimeters [8 inches] beneath the body of the whole ship.[104] [See the photo insert.]

---

[104] GEPAN, *Note Technique No. 16, Enquête 81/01, Analyse d'une Trace,* March 1, 1983.

The samples of soil and wild alfalfa collected from the landing site, as well as the control samples from varying distances from the epicenter, were subjected to a number of analyses: physico-chemical analysis at the SNEAP laboratory, electronic diffraction studies at Toulouse University, mass spectrometry by ion bombardment at the University of Metz, and biochemical analysis of the vegetable samples at the National Institute of Agronomy Research (INRA), among others.[105]

The Trans-en-Provence case is very likely the most thoroughly scientifically documented CE-II (Close Encounter of the Second Kind) ever investigated. Some of the scientific findings included:

> *Traces were still perceptible 40 days after the event.*
>
> *There was a strong mechanical pressure forced (probably the result of a heavy weight) on the surface.*
>
> *A thermatic heating of the soil, perhaps consecutive to or immediately following the shock, the value of which did not exceed 600 degrees.*
>
> *The chlorophyll pigment in the leaf samples was weakened from 30 to 50 percent . . . The young leaves withstood the most serious losses, evolving toward the content and composition more characteristic of old leaves.*
>
> *The action of nuclear irradiation does not seem to be analogous with the energy source implied with the observed phenomenon; on the other hand, a specific intensification of the transformation of chlorophyll . . . could be tied to the action of a type of electric energy field.*
>
> *On the biochemical level, the analysis was made*

---

[105] Velasco, Jean-Jacques, "Report on the Analysis of Anomalous Physical Traces: The 1981 Trans-en-Provence UFO Case," *Journal of Scientific Exploration*, vol. 4, no. 1, 1990.

on the entirety of the factors of photosynthesis, lipids, sugars and amino acids. There were many differences between those samples further from the spot of the landing and those that were closer to the spot.

It was possible to qualitatively show the occurrence of an important event which brought with it deformations of the terrain caused by mass, mechanics, a heating effect, and perhaps certain transformations and deposits of trace minerals.

We cannot give a precise and unique interpretation to this remarkable combination of results. We can state that there is, nonetheless, another confirmation of a very significant event which happened on this spot.[106]

Most of the puzzling biochemical mutations were discovered by Michel Bounias of INRA. Describing the young leaves to a journalist from *France-Soir* magazine, Bounias stated in 1983:

From an anatomical and physiological point, they [leaves] had all the characteristics of their age, but they presented the biochemical characteristics of leaves of an advanced age: old leaves! And that doesn't resemble anything that we know on our planet.[107]

In a technical report published in the *Journal of Scientific Exploration*, Bounias concluded:

It was not the aim of the author to identify the exact nature of the phenomenon observed on the 8th of January

---

[106] GEPAN, ibid.

[107] Roussel, Robert, *Les Vérités Cachées de l'Enquête Officielle*, Albin Michel, 1994; quoted in Huneeus, A., "The French government UFO dossier," *Fate*, October 1994.

Summary and Conclusion of Trans-en-Provence Case

Witness	No particular expectations	
Account	Precise coherent	
Psychological environment	No special influence	
Physical environment environment	-Impact -Friction	Iron, Zinc deposit temperature <600°
Biological effects noted	Premature aging of young alfalfa plants	-Photosynthesis modified -Time/space effect

Conclusion

-Physical phenomenon of unexplained nature
-High probability of electromagnetic mode of propulsion

Chart of Summary and Conclusion of Trans-en-Provence case. Courtesy of CNES/SEPRA.

1981 at Trans-en-Provence. But it can reasonably be concluded that something unusual did occur that might be consistent, for instance, with an electromagnetic source of stress. The most striking coincidence is that at the same time, French physicist J. P. Petit was plotting the equations that led, a few years later (Petit, 1986), to the evidence that flying objects could be propelled at very high speeds without turbulence nor shock waves using the magnetohydrodynamic effects of Laplace force action![108] [See box above.]

---

[108] Bounias, Michel, "Biochemical Traumatology as a Potent Tool for Identifying Actual Stresses Elicited by Unidentified Sources: Evi-

Out of a total of twenty-five hundred reports collected officially in France since 1977 and investigated by GEPAN, this case and three other ground-trace incidents (where strange ground traces were left after alleged UFO landings) continue to puzzle the original investigator, Jean-Jacques Velasco. At a meeting of the Society for Scientific Exploration (SSE) in Glasgow in 1994, Velasco summarized the "four noteworthy cases" with "effects observed on vegetation" (see page 119):

These cases have all been the subject of enquiries by the police, then GEPAN or SEPRA. In each of these situations, a UAP [Unidentified Aerospace Phenomena] was observed in direct relation in a zone perturbed by the phenomenon.

1. "Christelle" case of 27/11/1979: Persistence of flattened grass several days after the observation. The samples taken and analyzed by a plant biology laboratory at Toulouse University did not give unequivocal evidence of chemical or biological disturbance of the samples taken from the marked area relative to controls. A study of the mechanical properties of grass tissue subjected to strong mechanical pressure showed that the duration is a more important factor than the mass.

2. "Trans-en-Provence" case of 8/01/81: Apparition of a circular print in a crown shape after observation of a metallic object resting on the ground. The vegetation, a kind of wild alfalfa, showed withering of the dried leaves in the central part of the print. The analyses revealed damage of a specific kind affecting the functional relationships of the photosynthetic system.

---

dence for Plant Metabolic Disorders in Correlation with a UFO Landing," *Journal of Scientific Exploration*, vol. 4, no. 1, 1990.

3. "Amarante" case of 21/08/82: Severe drying of the stems and leaves on a bush (amaranth), punctuated by the appearance of raised blades of grass before the phenomenon disappeared. Biochemical analyses revealed that no reported outside agent could be the cause of such effects. Only a corona effect due to powerful electromagnetic fields could partially explain the observations.

4. "Joe Le Taxi" case of 7/09/87: Leaf damage on a tree (birch) and functional disturbance of the photosynthetic system after an intense light and sound phenomenon had been observed. This case demonstrated the importance of good sample collection and preservation for biochemical analysis.[109]

Of these four cases, Trans-en-Provence still remains the best documented. Velasco concluded that, after years of investigations:

The laboratory conclusion that seems to best cover the effects observed and analyzed is that of a powerful emission of electromagnetic fields, pulsed or not, in the microwave frequency range.[110]

SEPRA's latest thrust in the investigation has centered on "experimentally reproducing in the laboratory continuous and pulsed emissions of microwave fields at various powers and frequencies so as to verify biochemical effects on plants." While the studies are still preliminary, Velasco concluded his SSE presentation with the following statement:

---

[109] Velasco, J-J., "Action Of Electromagnetic Fields In The Microwave Range On Vegetation," paper presented at a meeting of the Society for Scientific Exploration in Glasgow, Scotland, August 1994.

[110] Ibid.

Painting of the Madonna and Saint Giovannino, in the Palazzo Vecchio in Florence, attributed to the 15th-century school of Filippo Lippi. A strange object can be glimpsed over the Madonna's left shoulder. A close-up **(inset)** of the upper right section of the painting shows an object in the sky, and below, a man and dog looking at the object. Enlargement shows this to be an oval or discoid craft with spikes of light painted around its perimeter. In other words, this is what today would be called a UFO. (Photographs courtesy of CUFOS)

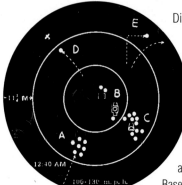

Diagram of the UFOs tracked by Washington's National Airport radar scope on July 20, 1952. At (A), 7 objects approach the nation's capital from the south. At (B), some are seen over the White House and Capitol. At (C), they appear over Andrews Air Force Base. At (D), one UFO tracks an airliner. At (E), one is seen to make a sharp right-angular turn. (Courtesy of *UFOs—A Pictorial History from Antiquity to the Present* by David C. Knight [McGraw Hill Book Co., 1979])

Enlargement of first UFO photograph taken by Almiro Barauna from a Brazilian Navy ship off Trindade Island. (Photo courtesy of the National Institute for Discovery Science)

Enlargement of second photograph taken by Almiro Barauna. (Photo courtesy of the National Institute for Discovery Science)

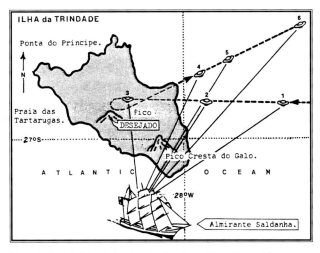

Map of Trindade Island showing the location of the ship and the trajectory of the UFO with the position where Barauna's photographs were taken. (Courtesy of the National Institute for Discovery Science)

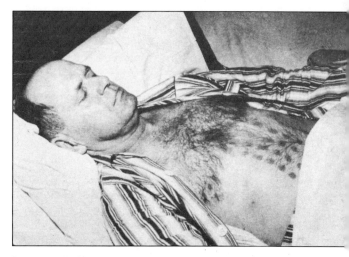

Photograph of Stephen Michalak showing the geometric pattern burned into his body by the exhaust gases of a UFO. (Courtesy of the National Institute for Discovery Science)

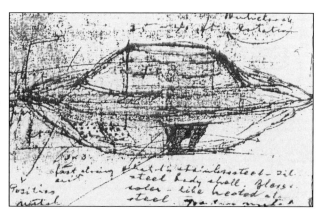

Drawing with notes by Michalak of the landed UFO he encountered at Falcon Lake. Notice grill pattern (encircled) and compare to burn marks above. From the declassified files of the RCAF and RCMP. (Courtesy of the National Institute for Discovery Science)

Photograph of ground traces deposited by a UFO that landed in Trans-en-Provence in January 1981. Analysis of the ground traces found odd electromagnetic stresses and evidence of accelerated aging. (Courtesy of CNES/SEPRA)

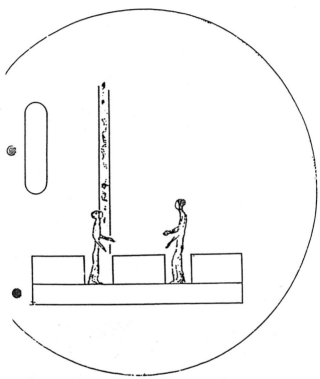

A drawing of a transparent sphere housing two tall occupants that was seen by a medical doctor and two witnesses over the Spanish Canary Islands. From the declassified file of the Spanish Air Force. (Courtesy of Antonio Huneeus)

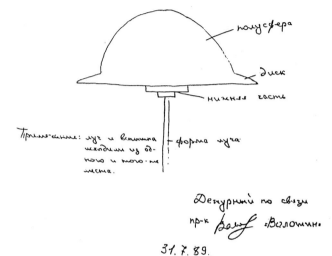

полусфера

диск

нижняя часть

Примечание: луч и величина исходили из од- ного и того-же места.

форма луча

Дежурный по связи пр-к *Волошин.*

31.7.89.

Drawing of a UFO (projecting some kind of beam) done by Ensign Voloshin after he witnessed unidentified craft fly over a Russian missile base in Kapustin Yar in July 1989. (Illustration courtesy of Antonio Huneeus/*Aura-Z*)

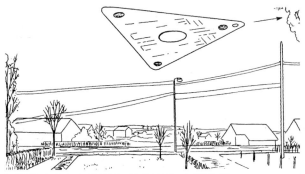

Fig. 1.46 - L'OVNI, tel qu'il fut observé par N.T., âgé de 12 ans, à Grand-Rechain, le 3 novembre 1993 (voir p. 108).

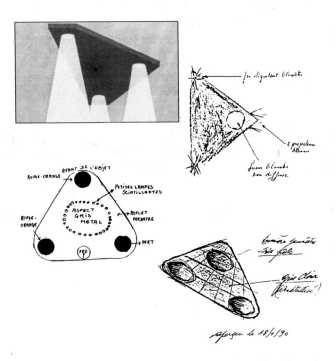

Montage of eyewitness drawings and reconstructions of triangular craft seen in Belgium between 1989 and 1993. (Courtesy of Guy Mossay, Artist Rights Society, NY, SOFAM, Brussels)

Cases with Physiological Signs

ENQUIRY CASE	VISIBLE SIGNS (PLANTS)	TYPE OF ANALYSIS	LABORATORY	SUGGESTED INTERPRETATION	STUDY
CHRISTELLE 27/11/79	Grass flattened in given direction	Plant physiology	UPS Toulouse Pr. TOUZE	None	Mechanical properties of grass tissue
TRANS-EN-PROVENCE 08/01/81	Alfalfa leaves withered	Biochemical	INRA Avignon Pr. BOUNIAS	Electromagnetic field Microwaves	Effect of microwaves on plants
AMARANTE 21/10/82	Grass raised Amaranth leaves withered, fruits burst	Plant physiology	UPS Toulouse Pr. ABRAVANEL	Electromagnetic field Microwaves	Effect of lightning on plants
JOE LE TAXI 07/09/4	Birch leaves damaged	Biochemical	INRA Pr. BOUNIAS	Electromagnetic field Microwaves	Effect of microwaves on plants

Chart of four UFO cases with physiological signs on vegetation. Courtesy of CNES/SEPRA.

*However these initial studies carried out to validate
the hypothesis of microwave action on the biological
activity of plants in relation with UAPs need to be extended if we are to understand the mechanisms involved at molecular scale. Similarly, an investigation
of the frequency range, the power and the exposure
time would be useful to confirm the hypothesis of
microwaves combined with other fields of electromagnetic forces coming into play in the propulsion of
UAPs.*[111]

---

[111] Ibid.

1986:
JET CHASE OVER BRAZIL

As many as twenty UFOs were seen and tracked by ground radar and at least six airplanes during the night of May 19, 1986, over several states in southeastern Brazil. Unidentified radar returns were tracked by airports in São Paulo and the Integrated Air Defense & Air Traffic Control Center (CINDACTA) in Brasilia. Two F-5E and three Mirage jet fighters were scrambled from Santa Cruz AFB in São Paulo State, and Anápolis AFB in Goias State.

The case was discussed openly by high-ranking government officials. It was first reported by Colonel (Ret.) Ozires Silva, president of the state-owned oil company Petrobrás, who was flying on an executive Xingu jet when he and the pilot saw and pursued unidentified objects for about twenty-five minutes. The incident was covered widely in the Brazilian media, leading to a press conference at the Ministry of Aeronautics in Brasilia on May 23, with air traffic controllers and Air Force pilots involved in the scramble mission.

The Minister of Aeronautics, Brigadier General Otávio Moreira Lima, was very outspoken:

Between 20:00 hrs. (5/19) and 01:00 hrs. (5/20) at least 20 objects were detected by Brazilian radars. They saturated the radars and interrupted traffic in the area. Each time that radar detected unidentified objects, fighters took off for intercept. Radar detects only solid metallic bodies and heavy (mass) clouds. There were no clouds nor conventional aircraft in the region. The sky was clear. Radar doesn't have optical illusions.

We can only give technical explanations and we don't have them. It would be very difficult for us

> *to talk about the hypothesis of an electronic war.*
> *It's very remote and it's not the case here in Brazil.*
> *It's fantastic. The signals on the radar were quite*
> *clear.*[112]

The Minister also announced that a commission would study the incident. Air Force Major Ney Cerqueira, in charge of the Air Defense Operations Center (CODA), was equally candid:

> *We don't have technical operational conditions to ex-*
> *plain it. The appearance and disappearance of these*
> *objects on the radar screens are unexplained. They*
> *are Unidentified Aerial Movements . . . The technical*
> *instruments used for the identification of the lights*
> *had problems in registering them. CODA activated*
> *two F-5E and three Mirages to identify the objects.*
> *One F-5E and one Mirage remained grounded on*
> *alert. A similar case occurred four years ago [the*
> *Commander Brito VASP airliner radar-visual inci-*
> *dent in 1982]. The lights were moving at a speed*
> *ranging between 250 and 1,500 km./hr. [150 to 1,000*
> *mph]. The Air Force has not closed the case.*[113]

Despite the initial openness shown by Brazilian authorities, the Ministry of Aeronautics Commission report was not released. However, the accounts of Air Force pilots and radar controllers were published widely in the press and later studied by Brazilian researchers. A comprehensive report was made by Basílio Baranoff, an airline captain, member

---

[112] Covo, Claudeir, "Maio de 86—A Mobilização No Céu Brasileiro," *o assunto é . . . UFOLOGIA,* no. 14, Editora Trés, São Paulo, 1986; contains transcripts of all the statements by military officers at the May 23, 1986, press conference.

[113] Ibid.

of the Aerospace Technical Center, and consultant for the Brazilian UFO organization CBPDV. Baranoff provided a chronology of events for the night of May 19, 1986:

6:30 P.M. local time—First visual sightings by control tower personnel at the São José dos Campos airport in São Paulo State. Controllers notice two intense lights aligned with the runway axis at 330 degrees azimuth and approximately 15 km. [10 m.] distance from the tower.

7:00 P.M.—The control towers in São Paulo and Brazilia confirm to São José that they have three primary targets on their screens, and that there are no scheduled aircraft in those areas.

8:00 P.M.—CINDACTA (Brasilia) detects up to eight targets (echoes) on its screens.

8:30 P.M.—A new object is observed with binoculars at the São José tower; it shows defined edges and red-orange color; it approaches the tower and then retreats.

9:00 P.M.—The oil company Xingu executive jet with Col. Silva requests landing conditions at São José. Both Silva and Commander Alcir Pereira, the Xingu pilot, confirm visually the luminous objects at 330 degrees azimuth. The Xingu jet attempts to follow the UFOs for 10 minutes.

9:10 P.M.—The Xingu jet returns for landing when a new, large luminous object heads toward the aircraft. The São Paulo tower confirms two echoes: the Xingu and an unknown, which disappears from the screen 15 minutes later.

9:20 P.M.—The Air Control Center in Brasilia (ACC-BR) informs the Air Defense Command about the situation.

9:25 P.M.—The Xingu returns for a second landing attempt when the São Paulo tower reports yet another object at 180 degrees south, which is observed and followed by Commander Pereira.

9:30 P.M.—The Xingu returns for a third landing attempt when ACC-BR reports the appearance of new objects. The Xingu, now at 3,000 meters [10,000 ft.] of altitude, makes visual contact with three luminous objects flying low over Petrobrás refineries and heading towards Serra do Mar. The Xingu finally lands in São José dos Campos.

9:40 P.M.—More visual sightings of a round object at 320 degrees azimuth.

9:50 P.M.—A luminous yellow object surrounded by smaller lights is observed at 110 degrees azimuth.

10:23 P.M.—The first F-5E jet fighter, piloted by Air Force Lt. Kleber Caldas Marinho, is scrambled from Santa Cruz AFB in Rio State.

10:45 P.M.—The second F-5E jet, piloted by Captain Brisola Jordão, is scrambled from Santa Cruz. The first Mirage F-103, armed with Sidewinder and Matra missiles, is scrambled from Anápolis AFB in Goias State.

10:55 P.M.—Anápolis AFB detects the objects on radar. The Mirage piloted by Captain Viriato does not make visual contact with the UFO, but a target is detected on its onboard radar . . . Captain Viriato later explained at the press conference in Brasilia that he was chas-

ing the UFO "at 1,350 km./hr. [850 mph], approaching the object up to a distance of 6 miles [9.5 km.]. The object was heading up front and moving from one side to the other (zigzagging) on my radar scope. Suddenly, the blip disappeared from my radar scope."

11:00 P.M.—The second Mirage F-103 is scrambled from Anápolis.

11:15 P.M.—Lt. Kleber's F-5E makes visual contact with a ball of light and chases the UFO at Mach 1.1 (1,320 km./hr. or 850 mph) . . . Kleber later declared at the press conference: "I had one visual contact and one contact with my aircraft radar of something that looked like a luminous point, which was 12 miles [19 km.] in front of me, a distance confirmed by ground radar. The object was moving from left to right and then began to climb . . . [It] was at 10 km [6 mi.] of altitude and flying over 1,000 km./hr. [600 mph]. I followed it up to 200 miles [320 km.] over the Atlantic Ocean [limit of Brazil's territorial waters]. I wasn't afraid because I like the unknown."

11:17 P.M.—The third Mirage jet is scrambled from Anápolis AFB.

11:20 P.M.—Captain Jordão's F-5E establishes radar contact . . . At the press conference a few days later, he stated: "Near to São José dos Campos, radar detected several targets, 10 to 13 targets, at a distance of 20 miles [32 km.]. The sky was clear but I didn't see anything. Ground radar informed me that the objects were closing in: 20 miles, 15, 10, 5, suddenly there were 13 objects behind my aircraft, 6 on one side and 7 on the other, during several minutes. After I maneuvered the aircraft, the objects had disappeared." [Captain Jordão flew for 1 hour 20 minutes.]

> *11:36 P.M.—The third Mirage is scrambled from Anápolis AFB.*

> *1:00 A.M. (May 20)—By this time all jet fighters have returned to their bases.[114]*

These are the basic known facts surrounding the multiple UFO jet-scramble incidents over southeastern Brazil on the night of May 19–20, 1986. It is noteworthy that Captain Baranoff added that "two nights after, ten to eleven unidentified luminous objects returned for a new round over São José dos Campos; they were observed visually and detected by the São Paulo, ACC-BR and by CINDACTA 1 radars." This time there was no official confirmation from Air Force authorities.[115]

Lacking the final Ministry Commission report with all the pertinent data, it is difficult to make a conclusion about this case. Many hypotheses were offered in the Brazilian media by skeptical astronomers and scientists, ranging from a meteor shower, a reflection of the full moon and ball lightning, to radar malfunction, space debris, and spy planes. Most of these explanations seem quite insufficient to explain the events of May 19. One of the more plausible was offered by British space researcher Geoffrey Perry. According to Perry, the Soviet space station Salyut-7 ejected several boxes of debris on that night, which reentered the Earth's atmosphere around central-western Brazil. The reentry of NASA's Solarwind satellite was also discussed in the Brazilian press.[116]

---

[114] Baranoff, Basílio, "Casuistica UFO—OVNIs em Maio de 1986," *PSI-UFO,* no. 4, CBPDV, Campo Grande, Jan./Feb. 1987. CBPDV stands for Centro Brasileiro para Pesquisas de Discos Voadores.

[115] Ibid.

[116] Huneeus, Antonio, "UFO Alert in Brazil," *UFO Review,* New York, 1986.

However, Brigadier José Cavalcanti, from Brazil's Air Defense Command, was not impressed with the Salyut-7 and Solarwind explanations. He told the weekly magazine *Veja*:

It could have been space debris, but it wasn't only that. A metallic box with space debris can be detected by radar, but it will always fall in the same direction and at constant velocity. That was not the case of what was seen in Brazil, where the objects detected by radar had speeds that varied from very slow to extremely high.[117]

Another interesting view is the final comment in a short message from the USDAO (U.S. Defense Attaché Office) in Rio to DIA (Defense Intelligence Agency) on the subject, entitled "BAF [Brazilian Air Force] has a Close Encounter of the First Kind":

COMMENT: [Censored] While RO [Reporting Officer] does not believe in UFO's or all the hoopla that surrounds previous reporting, there is too much here to be ignored. Three visual sightings and positive radar contact from three different types of radar systems, leads one to believe that something arrived over Brazil the night of 19 May.[118]

---

[117] *Veja*, "Lixo espacial—Surgem novas explicações para os OVNIs da FAB," September 10, 1986; English translation in Huneeus, A., ibid.

[118] Department of Defense JCS [Joint Chiefs of Staff] Message Center, "Info Report" re "Subject: BAF has a Close Encounter of the First Kind," May 1986.

1986:
JAPAN AIRLINES 747 OVER ALASKA

Japan Air Lines Flight 1628 was near the end of the Iceland–Anchorage leg of its flight from Paris to Tokyo with a cargo of wine, when its flight crew saw and tracked three unidentified objects. On the night of November 17, 1986, the sighting of at least one of the UFOs was initially confirmed by FAA and U.S. military ground radar.

According to Captain Kenju Terauchi, First Officer Takanori Tamefuji, and Flight Engineer Yoshio Tsukuda, two small lights and one huge lighted object were in sight on their radar for more than a half hour. They watched as they flew 350 miles (550 km.) southward across Alaska, from Fort Yukon toward Anchorage.

Captain Terauchi, a veteran of twenty-nine years' flying, said: "It was a very big one—two times bigger than an aircraft carrier." He changed altitude and made turns, with FAA permission, in an effort to identify the objects, which continued to follow him. He said the objects moved quickly and stopped suddenly. At one time, the light from the large object was so bright that it lit the airplane's cockpit and Captain Terauchi said he could feel heat from it on his face. He added that he had been watching the UFO for six minutes before notifying anyone on the ground; this would make the start of the sighting about 6:13 P.M.

The FAA at first confirmed the claims that several of its radar traffic controllers tracked the 747 and the large object, and that U.S. Air Force radar did as well. Later official statements hedged on this, and tried to ascribe the radar targets to weather effects. At the end, however, an FAA spokesman stated, "We are accepting the descriptions of the crew, but are unable to support what they saw."[119]

---

[119] Federal Aviation Administration (FAA) report, December 29, 1986.

The summary of the communication between JAL Captain Terauchi and ground controllers was published by the Federal Aviation Administration:

> *6:19 P.M. local time—The pilot of JL1628 requested traffic information from the ZAN (FAA Air Route Traffic Control Center, Anchorage) Sector 15 controller.*

> *6:26 P.M.—ZAN contacted the Military Regional Operations Control Center (ROCC), and asked if they were receiving any radar returns near the position of JL1628. The ROCC advised that they were receiving a primary radar return in JL1628's 10 o'clock (left-front) position at 8 miles [13 km.].*

> *6:27 P.M.—The ROCC contacted ZAN to advise they were no longer receiving any radar returns in the vicinity of JL1628.*

> *6:31 P.M.—JL1628 advised that the "plane" was "quite big," at which time the ZAN controller approved any course deviations needed to avoid the traffic.*

> *6:32 P.M.—JL1628 requested and received a descent from FL350 to FL310 (flight level 350 and 310, meaning altitude of 35,000 and 31,000 feet, or 10,500 m. and 9,500 m.). When asked if the traffic was descending also, the pilot stated it was descending "in formation."*

> *6:35 P.M.—JL1628 requested and received a heading change to two one zero (210 degrees, or southwest). The aircraft was now in the vicinity of Fairbanks and ZAN contacted Fairbanks Approach Control asking if they had any radar returns near JL1628's position. The Fairbanks Controller advised they did not.*

6:36 P.M.—JL1628 was issued a 360 degree turn and asked to inform ZAN if the traffic stayed with them.

6:38 P.M.—The ROCC called ZAN advising they had confirmed a "flight of two" in JL1628's position. They advised they had some "other equipment watching this," and one was a primary target only.

6:39 P.M.—JL1628 told ZAN they no longer had the traffic in sight.

6:42 P.M.—The ROCC advised it looked as though the traffic had dropped back and to the right of JL1628, however, they were no longer tracking it.

6:44 P.M.—JL1628 advised the traffic was now at 9 o'clock (left).

6:45 P.M.—ZAN issued a 10 degree turn to a northbound United Airlines flight, after pilot concurrence, in an attempt to confirm the traffic.

6:48 P.M.—JL1628 told ZAN the traffic was now at 7 o'clock (left rear), 8 miles [13 km.].

6:50 P.M.—The northbound United flight advised they had the Japan Airlines flight in sight, against a light background, and could not see any other traffic.

6:53 P.M.—JL1628 advised that they no longer had contact with the traffic.[120]

---

[120] Federal Aviation Administration (FAA), "Chronological Summary of the Alleged Aircraft Sightings by Japan Airlines Flight 1628," January 6, 1987.

Official statements became increasingly negative as the days passed, casting doubt on the radar confirmation of the visual observations. But at a press conference held by the FAA on March 5, spokesman Paul Steucke stated: "As far as we know, the whole crew are people of integrity and did report what they saw accurately."[121]

The JAL case was analyzed by optical physicist Dr. Bruce Maccabee, who divided it into four phases. In the first phase, Captain Terauchi saw some distant lights below and to his left, which seemed to pace the 747. Terauchi initially thought they were military aircraft, but was told by the Control Center that there was no traffic in the area. The second phase was the multiple-witness sighting by the whole crew when the lights moved abruptly in front of the aircraft. Maccabee wrote:

The multiple-witness sighting of the arrays of lights seems inexplicable. The sighting cannot have been a hallucination by the whole crew. The lights could not have been stars or planets. These stars and planets were visible before the "ships" appeared in front of the plane and were still visible after the "ships" had moved away. There is no natural phenomenon that can account for the rectangular arrangement of lights in horizontal rows, for the occasional sparking, for the vertical rectangular dark space between the rows, for the reorientation of the pairs of arrays from one above the other to one beside the other, for the heat which the captain felt on his face, and so on.[122]

---

[121] Statement by FAA Regional Director Paul Steucke, at March 5, 1987, press conference in Anchorage, Alaska. See Anchorage *Daily News*, March 6, 1987.

[122] Maccabee, Bruce, "The Fantastic Flight of JAL1628," *International UFO Reporter,* vol. 12, no. 2, CUFOS, March/April 1987.

The third phase occurred as the "ships" receded and could only be seen as "two dim, pale, white lights," but an echo was picked up by the aircraft onboard radar. While Maccabee concedes that the echo could have been caused by "a temporary (self-repairing) failure" (as suggested by the FAA), he added that "it seems much more likely that there was some object out there."

The last phase is perhaps the most controversial one, as Captain Terauchi was the only witness of the so-called giant spaceship "two times bigger than an aircraft carrier." Maccabee conceded:

> It seems at least plausible that he may have mis-interpreted oddly lighted clouds which the crew had reported to be below the aircraft. Although the several ground radar returns behind the jet were intriguing, the failure of the radar to show a continuous track of some unknown primary target makes the radar con-firmation ambiguous at best. Therefore it seems that, at the very least, the last portion of the sighting is not so convincing as the earlier portions.
>
> Even if one arbitrarily ignores that latter part of the "Fantastic Flight of JAL1628" one is still left with an intriguing sighting of the two "ships" which paced the aircraft. It seems, then, that the JAL1628 was accompanied during part of its flight by at least two TRUFOS (True UFOs).[123]

---

[123] Ibid.

1989: MULTIPLE WITNESS CASE AT RUSSIAN MISSILE BASE

Very little was known about UFO investigations in Russia and the republics of the former USSR during the communist era. UFOs were officially labeled "capitalist propaganda" in the 1950s and 1960s. A few scientists such as Professor Felix Zigel and Yuri Fomin documented UFO incidents, but their results were rarely published and circulated mostly in *samizdat* (unofficial) form. Ufology began to prosper in the early 1980s when "Commissions on Anomalous Phenomena" were established under the patronage of a few academicians.

Stories of secret military UFO investigations began to spread with *glasnost*, increasing with the breakup of the USSR. Retired military and intelligence officers were now speaking up and offering documents. One collection, covering military UFO investigations from 1978 to 1988, was sold by its former director, Colonel Boris Sokolov, to American journalist George Knapp and to ABC News.[124] In 1991, the Committee of State Security (KGB) declassified 124 pages of documents of "Cases of Observations of Anomalous Occurrences in the Territory of the USSR, 1982–1990," covering a total of 17 regions.[125]

One of the most interesting cases in the KGB file is a multiple-witness CE-I (Close Encounter of the First Kind) at an Army missile base in the district of Kapustin Yar, Astrakhan Region, on the night of July 28–29, 1989. The file is

---

[124] Knapp, George, "What the Russians Know About UFOs," *MUFON 1994 International UFO Proceedings;* "KGB UFO Files," ABC News *Prime Time Live*, broadcast originally in October 1994.

[125] KGB UFO File released in 1991; excerpts of the documents published in English in "Classification: 'Secret'—From the KGB Archives," *AURA-Z*, no. 1, Moscow, March 1993.

surely incomplete, but still offers an interesting glimpse into the maneuverability of UFOs. The dossier consists of the depositions of seven military witnesses (two junior officers, a corporal, and four privates) plus illustrations of the object by the observers, and a brief case summary by an unnamed KGB officer. (Neither the author nor the department is identified, but the document is at the beginning of the KGB file on the Kapustin Yar incident). It states in part:

> *Military personnel of the signal center observed UFOs in the period from 22:12 hrs. to 23:55 hrs. on 28 July 1989. According to the witnesses' reports, they observed three objects simultaneously, at a distance of 3–5 km. [2–3 mi.].*

A nearby base reported the UFO from 23:30 hrs. on July 28 until 1:30 hrs. on July 29. The report continues:

> *After questioning the witnesses, it was determined that the reported characteristics of the observed UFOs are: disc 4–5 m. [13–17 ft.] diameter, with a half-sphere on top, which is lit brightly. It moved sometimes abruptly, but noiselessly, at times coming down and hovering over ground at an altitude of 20–60 m. [65–200 ft.]. The command of [censored] called for a fighter . . . but it was not able to see it in detail, because the UFO did not let the aircraft come near it, evading it. Atmospheric conditions were suitable for visual observations.[126]*

The KGB file on the case is obviously incomplete, since there is no data on the jet scramble mission or whether

---

[126] KGB file entitled "Communication on Observation of Anomalous Event in the District of Kapustin Yar (July 28, 1989)"; English translation by Dimitri Ossipov.

ground or airborne radar detection was also reported. Nevertheless, the handwritten descriptions by the seven witnesses from the signal center do provide interesting reading about the flight behavior exhibited by the UFOs. The most detailed communication was submitted by the Officer-on-Duty, Ensign Valery N. Voloshin. A captain from the telegraph center informed him at 23:20 hrs. that "an unidentified flying object, which he called a flying saucer, was hovering over the military unit for over an hour." After confirming the sighting with the Operation Signal Officer on duty, Ensign Voloshin and Private Tishchayev climbed the first part of an antenna tower. According to his deposition:

One could clearly see a powerful blinking signal which resembled a camera flash in the night sky. The object flew over the unit's logistics yard and moved in the direction of the rocket weapons depot, 300 meters [1,000 ft.] away. It hovered over the depot at a height of 20 meters [65 ft.]. The UFO's hull shone with a dim green light which looked like phosphorous. It was a disc, 4 or 5 m. [13–17 ft.] in diameter, with a semispherical top.

While the object was hovering over the depot, a bright beam appeared from the bottom of the disc, where the flash had been before, and made two or three circles, lighting the corner of one of the buildings . . . The movement of the beam lasted for several seconds, then the beam disappeared and the object, still flashing, moved in the direction of the railway station. After that, I observed the object hovering over the logistics yard, railway station and cement factory. Then it returned to the rocket weapons depot, and hovered over it at an altitude of 60–70 m. [200–240 ft.]. The object was observed from that time on, by the first guard-shift and its commander. At 1:30 hrs., the object flew in the direction of the city of

*Akhtubinsk and disappeared from sight. The flashes
on the object were not periodical, I observed all this
for exactly two hours: from 23:30 to 1:30.*[127]

A drawing of the UFO was attached (see the photo insert.).
Private Tishchayev essentially confirmed Ensign Voloshin's testimony. The guard-shift of Corporal Levin and
Privates Bashev, Kulik, and Litvinov basically tell the same
story. They were all alerted by 1st Lt. Klimenko and they all
saw up to three UFOs performing fantastic acrobatics in the
sky, such as:

*Suddenly, it flew in our direction. It approached fast
and increased in size. It then like divided itself in
three shining points and took the shape of a triangle.
Then it changed course and went on flying in the
same sector.*[128]

> *After veering, it began to approach us and its
speed could be felt physically. (It swelled in front of
our eyes). Its flight was strange: no aircraft could fly
in this manner. It could instantly stop in the air (and
there was an impression that it wobbled slightly up
and down); it could float (exactly that: float, because
the word "fly" would not be adequate, it was as if the
air was holding it, preventing it from falling). At all
times that I observed it, it was blinking, blinking without any order and constantly changing colors (red,
blue, green, yellow). The point itself was not blinking
but something above it.*[129]

> *Here is what I observed: there was a flying object,
resembling an egg, but flatter. It shone brightly alter-
nating green and red lights. This object gathered a*

---

[127] Ensign Voloshin's Report in the Kapustin Yar KGB file, ibid.

[128] Deposition by Pvt. Bashev, Kapustin Yar KGB file, ibid.

[129] Deposition by Cpl. Levin, Kapustin Yar KGB file, ibid.

> *great speed. It accelerated abruptly and also stopped*
> *abruptly, all the while doing large jumps up or down.*
> *Then appeared a second and then a third object. One*
> *object rose to low altitude and stopped. It stayed there*
> *in one place and was gone. Later a second object dis-*
> *appeared, and only one stayed. It moved constantly*
> *along the horizon. At times, it seemed it landed on the*
> *ground, then it rose again and moved.*[130]

All the testimonies coincided with the appearance of a jet fighter attempting to intercept the UFOs. The fighter made a first pass above the object apparently without seeing it. Then, according to Lt. Klimenko's deposition, "the airplane, which could be identified by its noise, approached the object, but the object disengaged so fast, that it seemed the plane stayed in one place."

It is difficult to make a final evaluation of the Kapustin Yar CE-I, since no information about the scramble mission and possible radar tracking was released by the KGB. But the detailed testimony of seven military witnesses, who were familiar with rocket launches and various aircraft because of their post (Kapustin Yar is somewhat equivalent to the White Sands Proving Grounds in New Mexico), appears to confirm the unusual flight characteristics and extraordinary maneuverability displayed by UFOs in many instances. Moreover, as in the SAC flap of 1975 and the Bentwaters affair in England in 1980, the UFOs seemed capable of "demonstrating a clear intent in the weapons storage area," as described in a 1975 declassified teletype concerning Loring AFB in Maine.[131]

One of the official milestones of Soviet/Russian Ufology occurred less than a year later, as a result of a radar-visual and jet scramble incident in the Pereslavl-Zalesskiy region,

---

[130] Deposition by Lt. Klimenko, Kapustin Yar KGB file, ibid.

[131] Fawcett, L., and Greenwood, B., ibid.

east of Moscow, on the night of March 21, 1990. A statement issued by Colonel-General of Aviation Igor Maltsev, Chief of the Main Staff of the Air Defense Forces, was published in the newspaper *Rabochaya Tribuna*. Unit commanders compiled "more than 100 visual observations" and passed them on to Gen. Maltsev, who stated:

> *I am not a specialist on UFOs and therefore I can only correlate the data and express my own supposition. According to the evidence of these eyewitnesses, the UFO is a disc with a diameter from 100 to 200 meters [320 to 650 feet]. Two pulsating lights were positioned on its sides . . . Moreover, the object rotated around its axis and performed an "S-turn" flight both in the horizontal and vertical planes. Next, the UFO hovered over the ground and then flew with a speed exceeding that of the modern jet fighter by 2 or 3 times . . . The objects flew at altitudes ranging from 100 to 7,000 m. [300 to 24,000 ft.]. The movement of the UFOs was not accompanied by sound of any kind, and was distinguished by its startling maneuverability. It seemed the UFOs were completely devoid of inertia. In other words, they had somehow "come to terms" with gravity. At the present time, terrestrial machines could hardly have any such capabilities.*[132]

---

[132] "UFOs on Air Defense Radars," Rabochaya Tribuna, Moscow, April 19, 1990; English translation by the U.S. Foreign Broadcast Information Service (FBIS).

1989–1990: UFO SIGHTING WAVE IN BELGIUM

From October 1989 throughout 1990, hundreds of reports of lighted objects, frequently described as enormous and triangular in shape, were recorded in Belgium. Air Force supersonic F-16 jets chased these strange objects, which were simultaneously tracked by both airborne and ground radars. The Belgian government cooperated fully with civilian UFO investigators, an action without precedent in the history of government involvement in this field.

The Chief of Operations of the Royal Belgian Air Force, Colonel Wilfred De Brouwer (now Major General and Deputy Chief of the Belgian Air Force), set up a Special Task Force Unit to work closely with the Gendarmerie to investigate the sightings as soon as they were reported. Among the thousands of witnesses were many military and police officers, pilots, scientists, and engineers. The wave was documented by the Belgian Society for the Study of Space Phenomena (SOBEPS), a private organization from Brussels, which published two thick volumes on the UFO wave.[133]

The first important case was a multiple-witness observation of a strange aircraft, reported by gendarmes on patrol near the town of Eupen, not far from the German border. Auguste Meessen, professor of physics at the Catholic University in Louvain and a scientific consultant of SOBEPS, summarized the case:

> *On November 29, 1989, a large craft with triangular shape flew over the town of Eupen. The gendarmes von Montigny and Nicol found it near the road linking*

---

[133] SOBEPS, *Vague d'OVNI sur la Belgique—Un Dossier Exceptionnel,* Brussels 1991; *Vague d'OVNI sur la Belgique 2—Une Enigme Non Résolue,* Brussels, 1994.

Aix-la-Chapelle and Eupen. It was stationary in the air, above a field which it illuminated with three power-ful beams. The beams emanated from large circular surfaces near the triangle's corners. In the center of the dark and flat understructure there was some kind of "red gyrating beacon." The object did not make any noise. When it began to move, the gendarmes headed towards a small road in the area over which they ex-pected the object to fly. Instead, it made a half-turn and continued slowly in the direction of Eupen, following the road at low altitude. It was seen by different wit-nesses as it flew above houses and near City Hall.[134]

Sightings continued to be logged by SOBEPS and the Gendarmerie during the fall and winter of 1989–1990. Most witnesses described seeing dark, triangular objects with white lights at the corners and a red flashing light in the mid-dle. Many of the objects were said to have hovered, with some of them then suddenly accelerating to a very high speed. Most of the objects made no sound, but some were said to have emitted a faint humming like that of an electric motor.

Public interest in the wave reached its peak with a radar-visual and jet scramble incident on the night of March 30–31, 1990. This scramble was seen and reported by hun-dreds of citizens. A preliminary report prepared by Major P. Lambrechts of the Belgian Air Force General Staff was re-leased to SOBEPS. The report includes a detailed chronol-ogy of events and dismisses several hypotheses such as optical illusions, balloons, meteorological inversions, mili-tary aircraft, holographic projections, etc.

The incident began at 22:50 hrs. on March 30 when the Gendarmerie telephoned the radar "master controller at Glons" to report "three unusual lights forming an equilateral

---

[134] Meessen, Auguste, "Observations, analyses et recherches," chap-ter 10 in *Vague d'OVNI 2*, ibid.

triangle." More gendarmes confirmed the lights in the following minutes. When the NATO facility at Semmerzake detected an unknown target at 23:49 hrs., a decision to scramble two F-16 fighters was made. The jets took off at 0:05 hrs. from Beauvechain, the nearest air base, on March 31 and flew for just over an hour. According to Major Lambrecht's report:

> *The aircraft had brief radar contacts on several occasions, [but the pilots] . . . at no time established visual contact with the UFOs . . . each time the pilots were able to secure a lock on one of the targets for a few seconds, there resulted a drastic change in the behavior of the detected targets . . . [During the first lock-on at 0:13 hrs.] their speed changed in a minimum of time from 150 to 970 knots [170 to 1,100 mph and 275 to 1,800 km./hr.] and from 9,000 to 5,000 feet [2,700 m. to 1,500 m.], returning then to 11,000 feet [3,300 m.] in order to change again to close to ground level.[135]*

The Electronic War Center (EWC) of the Air Force undertook a much more detailed technical analysis of the F-16 computerized radar tapes, led by Colonel Salmon and physicist M. Gilmard. Their study was completed in 1992 and was later reviewed by Professor Meessen.

Although many aspects of this case remain unexplained, Meessen and SOBEPS have basically accepted the Gilmard-Salmon hypothesis that *some* of the radar contacts were really "angels" caused by a rare meteorological phenomenon. This became evident in four lock-ons, "where the object descended to the ground with calculations showing *negative* [emphasis added] altitude . . . It was evidently impossible

---

[135] Lambrechts, Major P., "Report Concerning the Observation of UFOs During the Night of March 30–31, 1990," preliminary report dated May 31, 1990.

that an object could penetrate the ground, but it was possible that the ground could act as a mirror." Meessen explained how the high velocities measured by the Doppler radar of the F-16 fighters might result from interference effects. He points out, however, that there is another radar trace for which there is no explanation. As for the visual sightings of this event by the gendarmes and others, Meessen suggests that they could possibly have been caused by stars seen under conditions of "exceptional atmospheric refraction."[136]

In a recent interview, Major General De Brouwer summarized his reflections on this complex case:

> *What impressed me the most were the witnesses, some of whom I know personally and convinced me that, in fact, something was going on. These were credible people and they told clearly what they saw.*
>
> *We always look for possibilities which can cause errors in the radar systems. We can not exclude that there was electromagnetic interference, but of course we can not exclude the possibility that there were objects in the air. On at least one occasion there was a correlation between the radar contacts of one ground radar and one F-16 fighter. This weakens the theory that all radar contacts were caused by electromagnetic interference. If we add all the possibilities, the question is still open, so there is no final answer.[137]*

The Belgian UFO wave yielded a rich volume of good-quality cases and many videos and photographs. One strikingly clear photograph of a triangular-shaped craft was taken at Petit-Rechain in early April 1990. As of 1994, it remained unexplained after numerous analyses, including a thorough computerized study at the Royal Military Academy.

---

136 Meessen, A., ibid.

137 Huneeus, A., telephone interview with Major General De Brouwer, October 5, 1995.

Although public interest in the Belgian wave reached its peak in the 1990–91 period, SOBEPS was still documenting cases in late 1993. Marc Valckenaer listed the main characteristics of the Belgian UFOs in the latest SOBEPS study. Various shapes such as round, rectangular, and cigar were reported, but the wave was dominated by triangular objects. Some of their characteristics included:

> *Irregular displacement (zigzag, instantaneous change of trajectory, etc.).*
> *Displacement following the contours of the terrain.*
> *Varying speeds of displacement (including very slow motion).*
> *Stationary flight (hovering).*
> *Overflight of urban and industrial centers.*
> *Sound effects (faint humming . . . to total silence).*[138]

Because the bulk of the Belgian sightings described triangular-shaped objects, many European and American researchers and journalists speculated that these were caused by either F-117A stealth fighters or some other revolutionary U.S. secret military aircraft. However, the only truly unusual characteristic of the F-117 is its near-invisibility to radar and infrared detection—it looks, flies, and sounds like any other subsonic jet airplane. Similar claims about the presence of other American advanced airplanes are even harder to substantiate: the A-12 Avenger II was never built, and the existence of the TR-3A "manta" is unconfirmed. Neither has even been rumored to be able to fly in the manner reported for the Belgian UFOs.

Despite the fact that the secret-military-aircraft hypothesis has been denied officially over and over again by the Belgian Ministry of Defense and Air Force, as well as by the U.S. Embassy in Brussels and the U.S. Defense Intelligence

---

[138] Valckenaer, Marc, "Etude des particularités remarquables," chapter 2 in *Vague d'OVNI 2*, ibid.

Agency, some publications continue to champion the stealth fighter theory.

In a letter to French researcher Renaud Marhic, the Minister of Defense at the time of the UFO wave, Leo Delcroix, wrote:

> *Unfortunately, no explanation has been found to date. The nature and origin of the phenomenon remain unknown. One theory can, however, be definitely dismissed since the Belgian Armed Forces have been positively assured by American authorities that there has never been any sort of American aerial test flight.*[139]

[139] Marhic, Renaud, "Ovnis belges: nouvelle rumeur," *Phénoména* no. 13, Jan./Feb. 1993, SOS OVNI, Aix-en-Provence, France.

1991–1994:
RECENT CASES

While this report discusses UFO cases only up through the Belgian wave of 1989–90, many impressive sightings have continued to be logged by both official and private entities around the world.

These cases have not been included because thorough investigations are still under way. In order to make a proper evaluation of the validity of a promising UFO case, an in-depth study with scientific scrutiny must be made, which requires a great deal of time. However, in order not to ignore some of these noteworthy "pending" cases, some of them are presented here:

1991: Paraguay. A radar-visual UFO sighting was reported on the night of June 8 from two airplanes: a private Cessna 210 carrying three passengers, and a Paraguayan Air Lines flight from Asunción to Miami. Air traffic control at Asunción's airport detected an unknown radar track and also saw an object hovering over a runway. An official document from the Civil Aeronautical Agency of the Ministry of Defense confirms the radar detection. The Cessna's automatic direction finder (ADF) malfunctioned during the sighting. Pilot César Escobar reported:

> *During close approaches, the Cessna instruments "went crazy" . . . The (ADF) was moving around indiscriminately, without any sense of direction. When the light moved a little farther away, everything returned to normal. It repeated this "game" several times. It seemed to be under intelligent control.*[140]

---

[140] Ramírez, Jorge A., "UFO Intercepts Aircraft Over Paraguay," *MUFON UFO Journal,* no. 310, February 1994.

1992–93: Mexico. In 1992–93, there were many UFO sightings over Mexico City. The sightings reported over the Benito Juárez International Airport on March 4–5, 1992, were confirmed by radar detection. One case was reported by pilots of two airliners while preparing to land around 4:00 P.M. on September 16, 1993. One pilot described the UFO as shaped like a praying mantis. He added:

> *It was a beautiful day. I first thought it was a balloon, but . . . it was going too fast. We saw it really good and it was not a plane.*[141]

1993: United Kingdom. Lights were widely reported moving erratically over Great Britain on the night of March 30–31 and were investigated by the Ministry of Defense. Five members of a family described a huge diamond-shaped object flying slowly over their heads with an unpleasant, low humming sound. An RAF meteorologist reported an object, at first stationary, then moving erratically toward him, with speeds of several hundred miles per hour. It then shone a beam of light toward the ground, which tracked across a field. He also heard a low humming sound.[142]

1994: United States. On the night of March 8, police in southwestern Michigan were flooded with calls from dozens of witnesses about strange lights and vague objects. Officer Jeff Velthouse was dispatched to the sighting area, where he confirmed the presence of three objects moving in the same direction. These objects were also recorded by a U.S. Weather Service radar. The radar operator commented:

---

[141] Maussán, Jaime, "OVNIS Sobre la Tierra," *Epoca,* Mexico, Nov. 15, 1993; quoted in Hunneus, Antonio, "UFO Chronicle: More UFOS and IFOs from Mexico (Part II)," *Fate,* December 1994.

[142] Pope, Nick, lecture at the New Hampshire MUFON Conference, Portsmouth, September 19, 1995.

There were three and sometimes four blips, and they weren't planes. Planes show as pin points on the scope; these were the size of half a thumb nail . . . They were moving all over the place. I never saw anything like it before, not even during severe weather.[143]

---

[143] Coyne, Shirley, "Michigan Visual & Radar Sightings," *MUFON UFO Journal*, no. 317, September 1994.

SUMMARY

These cases are among the most detailed, best authenticated, and most puzzling of the many thousands of unexplained UFO reports. They are by no means the only such cases. In fact, it is the great mass of baffling reports from expert witnesses, past and present, that form the basis of the UFO mystery.

When studied as a group, these case histories exhibit clear patterns that strongly suggest that they belong to a distinct new class of phenomena, rather than being a formless collection of disparate observational errors.

Each of the cases detailed here is representative of one or more characteristics of UFO reports: radar-visual detection, physical traces, air-to-air sightings, attempted intercepts, multiple witness observations, etc. Each of these characteristics can be found in dozens of other well-authenticated multiple witness cases and reports. Most of these cases, as well as many hundreds more, involve some degree of government activity.

The primary question remains: If UFOs are so different from all known phenomena, what are they? The great majority of sightings reported as UFOs can be explained as IFOs, but a significant percentage cannot, and it is those which constitute the mystery.

Some of them may be secret aircraft, since at any time there are legitimately classified projects being conducted by several governments. Others may be unknown natural phenomena, since there is no way to completely rule out things that are, by definition, unknown. Still others may well be hoaxes, though these have played, statistically, a minor role in the history of UFOs.

But this still leaves thousands of highly detailed descriptions of apparently manufactured devices that are capable of speed and maneuverability far in excess of anything known to have been built in the 1990s, let alone the 1940s.

It is this large quantity of evidence of the existence of something completely baffling that motivates many of us to urge the governments of the world to release all they know about UFOs so that the people of the world, and especially scientists, can begin to come to grips with a mystery that has for too long been subjected to secrecy and ridicule.

PART 3
QUOTATIONS

from
prominent government and military officials,
astronauts, and scientists
throughout the world

GOVERNMENTS

United States

Military / Intelligence

General Nathan D. Twining, Chairman of the Joint Chiefs of Staff (1957–60). As Lieutenant General in charge of the Air Force Air Materiel Command at Wright Field, Ohio, he reported in 1947 on his investigation of UFO sightings to date:

a. The phenomena reported is something real and not visionary or fictitious.

b. There are objects probably approximating the shape of a disc, of such appreciable size as to appear to be as large as a man-made aircraft.

c. There is a possibility that some of the incidents may be caused by natural phenomena, such as meteors.

d. The reported operating characteristics such as extreme rates of climb, maneuverability (particularly in roll), and action which must be considered evasive when sighted or contacted by friendly aircraft and radar, lend belief to the possibility that some of the objects are controlled either manually, automatically, or remotely. [Letter to the Commanding General of the U.S. Army Air Forces, September 23, 1947.]

J. Edgar Hoover, in response to a government request to study UFOs:

*I would do it, but before agreeing to do it, we must in-
sist upon full access to discs recovered. For instance,
in the L.A. [or La.] case, the Army grabbed it and
would not let us have it for cursory examination.*
[Handwritten note to Clyde Tolson, July 15, 1947.]

General Walter Bedell Smith, Director of the CIA from
1950 to 1953, stated:

*The Central Intelligence Agency has reviewed the
current situation concerning unidentified flying objects which have created extensive speculation in the
press and have been the subject of concern to Government organizations . . . Since 1947, approximately
2,000 official reports of sightings have been received
and of these, about 20% are as yet unexplained.*

*It is my view that this situation has possible impli-
cations for our national security which transcend the
interests of a single service. A broader, coordinated
effort should be initiated to develop a firm scientific
understanding of the several phenomena which apparently are involved in these reports . . .* [1952
memorandum to the National Security Council.]

General Douglas MacArthur:

*Because of the developments of science, all the coun-
tries on earth will have to unite to survive and to
make a common front against attack by people from
other planets. The politics of the future will be cosmic, or interplanetary.* [*The New York Times,* October 8, 1955.]

*You now face a new world—a world of change.
The thrust into outer space of the satellite, spheres
and missiles marked the beginning of another epoch
in the long story of mankind—the chapter of the
space age . . . We speak in strange terms: of harness-*

ing the cosmic energy . . . of the primary target in war,
no longer limited to the armed forces of an enemy, but
instead to include his civil populations; of ultimate
conflict between a united human race and the sinister
forces of some other planetary galaxy . . . [Address by
General Douglas MacArthur to the United States Military Academy at West Point, May 12, 1962.]

Captain Edward J. Ruppelt, Chief of Project Blue Book,
from his book, *The Report on Unidentified Flying Objects,* 1956:

Every time I get skeptical, I think of the other reports
made by experienced pilots and radar operators, sci-
entists, and other people who know what they are
looking at. These reports were thoroughly investi-
gated and they are still unknowns.

We have no aircraft on this earth that can at will
so handily outdistance our latest jets . . . The pilots,
radar specialists, generals, industrialists, scientists,
and the man on the street who have told me, "I
wouldn't have believed it either if I hadn't seen it my-
self," knew what they were talking about. Maybe the
Earth is being visited by interplanetary space ships.

His comments on the Lubbock lights case:

When four college professors, a geologist, a chemist,
a physicist and a petroleum engineer report seeing
the same UFOs on fourteen different occasions, the
event can be classified as, at least, unusual. Add the
fact that hundreds of other people saw these UFOs
and that they were photographed, and the story gets
even better. Add a few more facts—that these UFOs
were picked up on radar and that a few people got a
close look at one of them, and the story begins to con-
vince even the most ardent skeptic. [Ruppelt, Edward J.,

The Report on Unidentified Flying Objects, New York: Doubleday, 1956.]

Admiral Roscoe Hillenkoetter, first Director of the CIA (1947–50). In 1957, he joined the Board of Governors of the National Investigations Committee on Aerial Phenomenon (NICAP), a UFO investigating group. In 1960, he stated:

> *Unknown objects are operating under intelligent control . . . It is imperative that we learn where UFOs come from and what their purpose is . . .* [Maccabee, Bruce, "What The Admiral Knew: UFO, MJ-12 and R. Hillenkoetter," *International UFO Reporter,* Nov./ Dec., 1986.]

He also recommended:

> *It is time for the truth to be brought out in open Congressional hearings. Behind the scenes high ranking Air Force officers are soberly concerned about the UFOs. But through official secrecy and ridicule, many citizens are led to believe the unknown flying objects are nonsense.* [Statement in a NICAP news release, February 27, 1960.]

General Curtis LeMay, Air Force Chief of Staff in his 1965 autobiography, *Mission With LeMay,* stated that although the bulk of UFO reports could be explained as conventional or natural phenomena, some could not:

> *We had a number of reports from reputable individuals (well-educated serious-minded folks, scientists and fliers) who surely saw something.*
> *Many of the mysteries might be explained away as weather balloons, stars, reflected lights, all sorts of odds and ends. I don't mean to say that, in the un-*

*closed and unexplained or unexplainable instances,
those were actually flying objects. All I can say is that
no natural phenomena could be found to account for
them . . . Repeat again: There were some cases we
could not explain. Never could."* [Statement from 1965
autobiography, *Mission with LeMay,* with MacKinlay
Kantor, New York: Doubleday, 1965.]

Major General E. B. LeBaily, USAF Director of Information:

*Many of the reports that cannot be explained have
come from intelligent and technically well-qualified
individuals whose integrity cannot be doubted.* [September 28, 1965, letter to USAF Scientific Advisory
Board requesting a review of the UFO project. Gillmor, Daniel S., ed., "Scientific Study of Unidentified
Flying Objects" (The Condon Report), New York
Times Books, 1969.]

General George S. Brown, USAF Chief of Staff, addressed
the appearance of UFOs during the Vietnam War at a press
conference:

*I don't know whether this story has ever been told or
not. They weren't called UFOs. They were called
enemy helicopters. And they were only seen at night
and they were only seen in certain places. They were
seen up around the DMZ [demilitarized zone] in the
early summer of '68. And this resulted in quite a little
battle. And in the course of this, an Australian destroyer took a hit and we never found any enemy, we
only found ourselves when this had all been sorted
out. And this caused some shooting there, and there
was no enemy at all involved but we always reacted.
Always after dark. The same thing happened up at
Pleiku at the Highlands in '69.* [Department of Defense

transcript of press conference in Illinois, October 16, 1973.]

Lt. Col. Lawrence J. Coyne, U.S. Army Reserve helicopter pilot with three thousand hours of flying time. He and three airmen had a close encounter with a UFO on the night of October 18, 1973, while flying in a U.S. Army Bell Huey utility helicopter in the vicinity of Mansfield, Ohio. Lt. Coyne described his experience at a United Nations UFO hearing in 1978:

> *With the aircraft under my control, I observed the red-lighted object closing upon the helicopter at the same altitude at a high rate of speed. It became apparent a mid-air collision was about to happen unless evasive action was taken.*
>
> *I looked out ahead of the helicopter and observed an aircraft I have never seen before. This craft positioned itself directly in front of the moving helicopter. This craft was 50 to 60 feet long with a grey metallic structure. On the front of this craft was a large steady bright red light. I could delineate where the red stopped on the structure of this craft because red was reflecting off the grey structure. The design of this craft was symmetrical in shape with a prominent aft indentation on the undercarriage. From this portion of the undercarriage, a green light, pyramid-shaped, emerged with the light initially in the trail position. This green light then swung 90 degrees, coming directly into the front windshield and lighting up the entire cockpit of the aircraft. All colors inside the cabin of the helicopter were absorbed by this green light. That includes the instrument panel lights on the aircraft.*
>
> *As a result of my experience, I am convinced this object was real and that these types of incidents should require a thorough investigation. It is my own per-*

*sonal opinion that worldwide procedures need to be
established to effectively study this phenomena through
an international cooperative effort. The establishment of a Transponder Code for aircraft flying worldwide is needed, to identify to ground controllers that
a pilot is indeed experiencing a UFO phenomena and
that pilot anxiety can be reduced to provide safe effective flying, knowing he is under radar control.*
[Statement to the Special Political Committee of the
United Nations, November 27, 1978.]

Victor Marchetti, Former CIA official:

*We have, indeed, been contacted—perhaps even
visited—by extraterrestrial beings, and the U.S. government, in collusion with the other national powers
of the earth, is determined to keep this information
from the general public.*

*The purpose of the international conspiracy is to
maintain a workable stability among the nations of
the world and for them, in turn, to retain institutional
control over their respective populations. Thus, for
these governments to admit that there are beings from
outer space . . . with mentalities and technological
capabilities obviously far superior to ours, could,
once fully perceived by the average person, erode the
foundations of the earth's traditional power struc-
ture. Political and legal systems, religions, economic
and social institutions could all soon become mean-
ingless in the mind of the public. The national oli-
garchical establishments, even civilization as we now
know it, could collapse into anarchy.*

*Such extreme conclusions are not necessarily valid,
but they probably accurately reflect the fears of the
"ruling classes" of the major nations, whose leaders
(particularly those in the intelligence business) have*

always advocated excessive governmental secrecy as being necessary to preserve "national security." [Marchetti, Victor: "How the CIA Views the UFO Phenomenon," *Second Look,* vol. 1, no. 7, Washington, D.C., May 1979.]

United States

Presidents

Harry S. Truman, at the time he was President, commented:

I can assure you that flying saucers, given that they exist, are not constructed by any power on earth. [April 4, 1950, White House Press Conference.]

President Gerald Ford, in a letter he sent as a congressman to L. Mendel Rivers, Chairman of the Armed Services Committee, on March 28, 1966:

No doubt, you have noted the recent flurry of newspaper stories about unidentified flying objects (UFOs). I have taken special interest in these accounts because many of the latest reported sightings have been in my home state of Michigan . . . Because I think there may be substance to some of these reports and because I believe the American people are entitled to a more thorough explanation than has been given them by the Air Force to date, I am proposing that either the Science and Astronautics Committee or the Armed Services Committee of the House, schedule hearings on the subject of UFOs and invite testimony from both the executive branch of the Government and some of the persons who claim to have seen

UFOs ... In the firm belief that the American public
deserves a better explanation than that thus far given
by the Air Force, I strongly recommend that there be
a committee investigation of the UFO phenomena. I
think we owe it to the people to establish credibility
regarding UFOs and to produce the greatest possible
enlightenment on this subject. [Committee on Armed
Services of the House of Representatives, Eighty-
ninth Congress, Second session, Hearing on Unidentified Flying Objects, April 5, 1966.]

President Jimmy Carter during his election campaign in
May 1976:

If I become President, I'll make every piece of information this country has about UFO sightings available to the public, and the scientists. I am convinced
that UFOs exist because I've seen one ... [The National Enquirer, June 8, 1976, "The Night I Saw a
UFO." Statement confirmed by White House special
assistant media liaison Jim Purks in an April 20, 1979,
letter.]

President Ronald Reagan was often quoted referring to the
possibility of an alien threat. Describing discussions held
privately with General Secretary Gorbachev, he stated:

... when you stop to think that we're all God's chil-
dren, wherever we may live in the world, I couldn't
help but say to him, just think how easy his task and
mine might be in these meetings that we held if suddenly there was a threat to this world from some other
species from another planet outside in the universe.
We'd forget all the little local differences that we have
between our countries and we would find out once
and for all that we really are all human beings here

on this earth together. [White House transcript of "Remarks of the President to Fallston High School Students and Faculty," December 4, 1985.]

In an address to the United Nations General Assembly in September 1987:

In our obsession with antagonisms of the moment, we often forget how much unites all the members of humanity. Perhaps we need some outside, universal threat to make us recognize this common bond. I occasionally think how quickly our differences worldwide would vanish if we were facing an alien threat from outside this world. [Speech to the United Nations General Assembly, Forty-second session, "Provisional Verbatim Record of the Fourth Meeting," September 21, 1987.]

United States

Congress

Representative John W. McCormack (D-Massachusetts), Speaker of the House, stated in a November 4, 1960, letter to Major Donald Keyhoe:

Some three years ago [1957], as chairman of the House Select Committee on Outer Space out of which came the recently established NASA, my Select Committee held executive sessions on the matter of "Unidentified Flying Objects." We could not get much information at that time, although it was pretty well established by some in our minds that there were some objects flying around in space that were un-

explainable. [Hall, Richard, *The UFO Evidence,* NICAP, 1964.]

Representative Jerry L. Pettis (R-California) stated in 1968 during the House Committee on Science and Astronautics hearing on UFOs:

> *Having spent a great deal of my life in the air, as a pilot . . . I know that many pilots . . . have seen phenomena that they could not explain. These men, most of whom have talked to me, have been very reticent to talk about this publicly, because of the ridicule that they were afraid would be heaped upon them . . . However, there is a phenomena here that isn't explained.* [U.S. House of Representatives, Ninetieth Congress, Second session, July 29, 1968.]

Senator Barry M. Goldwater, Sr. (R-Arizona), Republican presidential candidate 1964. In a letter to researcher Shlomo Arnon, dated March 28, 1975, he stated:

> *The subject of UFOs is one that has interested me for some long time. About ten or twelve years ago, I made an effort to find out what was in the building at Wright Patterson Air Force Base where the information is stored that has been collected by the Air Force, and I was understandably denied the request. It is still classified above Top Secret. I have, however, heard that there is a plan under way to release some, if not all, of this material in the near future. I'm just as anxious to see this material as you are, and I hope we will not have to wait too much longer.* [Good, T., *Above Top Secret,* New York: Quill William Morrow, 1988; Frontispiece.]

In an April 11, 1979, letter to Mr. Lee M. Graham, he added:

It is true I was denied access to a facility at Wright Patterson. Because I never got in, I can't tell you what was inside. We both know about the rumors.

Apart from that, let me make my position clear: I do not believe that we are the only planet of some two billion that exist that has life on it. I have never seen what I would call a UFO, but I have intelligent friends who have, so I can sort of argue either way."
[Ibid.]

Representative Steven H. Schiff (R-New Mexico), in response to inquiries from his constituents in 1993 concerning a possible cover-up of the crash of an alleged UFO outside Roswell, New Mexico, in 1947, requested information from the Department of Defense. In a CBS radio interview in February 1994, he stated:

I wrote to the Dept. of Defense, laying out these allegations and asking them if someone could come over with the file and brief me on it. My intent was to simply release this back to whomever inquired, which is very routine in Congress.

The response I got was not routine. The response I got was a very brief letter from the Air Force saying that my request had been referred to the National Archives, without any further comment . . . and without any offer of any kind of assistance in retrieving it . . . So I went to the National Archives and the National Archives wrote a letter back to me saying they didn't have anything in their files on the Roswell incident . . . I just have to say this much: the way the Dept. of Defense has responded has not been routine.

Having been given a "runaround" in his search, he instigated an inquiry by the GAO (General Accounting Office) in 1994 into the handling of Air Force files relating to this matter.

> *I did not ask the General Accounting Office to try once and for all to resolve this matter . . . What I asked the GAO to do was to assist me in locating whatever Air Force and Defense Department files would have existed on the subject, or an accounting of what happened to them.*
>
> *To me the issue is government accountability. I think that people who want to see government records are entitled to see government records or to get an explanation of what happened to them, regardless of their reason, regardless of the subject matter. It was my intention simply to make that information public if I could . . . unless there is a present security reason why not—and I have to add real fast if the matter is classified "military secret," we members of Congress can't just go monkeying around in there anytime we want. There are procedures for us too and that's fine with me.*
>
> *I was not told that we have a file that's classified. I was simply referred to an agency which I have to believe—now that I know the prominence of the Roswell incident—I have to believe the Dept. of Defense knew very well that I wasn't going to find anything in the National Archives when they sent me there twice.*
>
> *It's difficult for me to understand even if there was a legitimate security concern in 1947, that it would be a present security concern these many years later. Frankly I am baffled by the lack of responsiveness on the part of the Defense Dept. on this one issue, I simply can't explain it.* [Excerpts of Congressman Schiff's remarks on CBS radio's *The Gil Gross Show*, February 1994.]

Argentina

Argentine Navy. In the 1960s, the Argentine Navy was charged with the official investigation of UFO sightings, particularly those reported by its own personnel. A 1965 "Official UFO Report" prepared by **Captain Sánchez Moreno** from the Naval Air Station Comandante Espora in Bahía Blanca revealed that:

> *Between 1950 and 1965, personnel of Argentina's Navy alone made 22 sightings of unidentified flying objects that were not airplanes, satellites, weather balloons or any type of known (aerial) vehicles. These 22 cases served as precedents for intensifying that investigation of the subject by the Navy. In the past two years, nine incidents have been recorded that are being studied by Captain Pagani and a team of military and civilian scientists and collaborators. Likewise, a meticulous questionnaire was drafted, printed and distributed to different bases. In a short time, the Service of Naval Intelligence was in possession of a stack of highly significant reports of testimonies. On the basis of this important documentation, it was possible to obtain a coherent overview of the problem.* [Captain Sánchez Moreno, *Informe Oficial O.V.N.I., Sumario S# A. 02778-DTO. OVNI,* Naval Air Station Comandante Espora, in ICUFON *Project World Authority for Spatial Affairs (W.A.S.A.),* New York, 1979.]

Captain Engineer Omar R. Pagani, Director of the Argentine Navy UFO investigation team in the 1960s. As a result of a series of observations at Argentine and Chilean meteorological stations on Deception Island, Antarctica, in June and July 1965, Captain Pagani disclosed at a press conference that:

The unidentified flying objects do exist. Their presence and intelligent displacement in the Argentine airspace has been proven. Their nature and origin is unknown and no judgement is made about them. [Sánchez, Moreno, ibid.]

In addition, the *Argentine Navy Bulletin #172* of July 7, 1965, reported:

From the Navy post at the South Orkney Islands comes a message of extreme importance: during the passage of the strange object over the base [earlier the same day], two magnetometers in perfect working condition registered sudden and strong disturbances of the magnetic field (at 17:03 hrs.), which were recorded on their tapes. [Perissé, Captain D. A., "Deception Island UFO Sightings," in the *MUFON 1987 International UFO Symposium Proceedings*, Washington, D.C., June 1987.]

Belgium

Major-General Wilfred De Brouwer, Deputy Chief, Royal Belgian Air Force:

In any case, the Air Force has arrived to the conclusion that a certain number of anomalous phenomena has been produced within Belgian airspace. The numerous testimonies of ground observations compiled in this [SOBEPS] book, reinforced by the reports of the night of March 30–31 [1990], have led us to face the hypothesis that a certain number of unauthorized aerial activities have taken place. Until now, not a single trace of aggressiveness has been signalled; military or civilian air traffic has not been perturbed or threatened. We can therefore advance that the presumed activities do not constitute a concrete menace.

The day will come undoubtedly when the phenomenon will be observed with technological means of detection and collection that won't leave a single doubt about its origin. This should lift a part of the veil that has covered the [UFO] mystery for a long time. A mystery that continues to the present. But it exists, it is real, and that in itself is an important conclusion. [De Brouwer, W., "Postface" in SOBEPS' *Vague d'OVNI sur la Belgique—Un Dossier Exceptionnel*, Brussels: SOBEPS, 1991.]

Brazil

Brigadier General João Adil Oliveira, Chief of the Air Force General Staff Information Service (with the rank of Colonel), led the first official military UFO inquiry in Brazil in the mid-1950s. In a "Briefing" to the Army War College in Rio de Janeiro on November 2, 1954, Col. Oliveira stated:

I wish to give you a summary of what is known in the world about "flying discs," of what is known about the opinion of qualified experts who have dealt with this matter. The problem of "flying discs" has polarized the attention of the whole world, but it's serious and it deserves to be treated seriously. Almost all the governments of the great powers are interested in it, dealing with it in a serious and confidential manner, due to its military interest. [Col. Oliveira's Briefing included short summaries of several UFO incidents in the United States and Brazil. The full text was published in *O'Cruzeiro* magazine, Rio de Janeiro, December 11, 1954; reprinted in Martins, João, as *Chaves do Mistério*, Rio: HUNOS 1979.]

Later promoted to the rank of Brigadier General, he was interviewed by the Brazilian press on February 28, 1958:

> *It is impossible to deny any more the existence of flying saucers at the present time . . . The flying saucer is not a ghost from another dimension or a mysterious dragon. It is a fact confirmed by material evidence. There are thousands of documents, photos, and sighting reports demonstrating its existence. For instance, when I went to the Air Force High command to discuss the flying saucers I called for ten witnesses— military (AF officers) and civilians—to report their evidence about the presence of flying saucers in the skies of Rio Grande do Sul, and over Gravataí AFB [Air Force Base]; some of them had seen UFOs with the naked eye, others with high powered optical instruments. For more than two hours the phenomenon was present in the sky, impressing the selected audience: officers, engineers, technicians, etc.* [How to doubt?" *O Globo*, Rio de Janeiro, February 28, 1958; cited in Fontes, Olavo, M.D., "UAO Sighting over Trindade," *The A.P.R.O. Bulletin*, May 1960.]

System of Investigation of Unidentified Aerial Objects (SIOANI). In 1969, the IV Aerial Zone in São Paulo (changed in 1973 to the IV Regional Aerial Command, IV COMAR) established a specialized UFO bureau called SIOANI, under Major (later Colonel) Gilberto Zani. A letter signed by **Colonel João Glaser** from the IV COMAR to the Brazilian UFO group CPDV, dated November 28, 1984, gives a summary of SIOANI's functions:

> *From 1969 to 1972, the Ufological activities of this organization (SIOANI) were most varied, including the elaboration of information bulletins, a draft of SIOANI regulations, contacts with interested parties, panels, catalogs of contacts and others, always attempting to*

contribute in this field of research that was already well known in Brazil. [Os Documentos Oficiais da Força Aérea Brasileira, Centro para Pesquisas de Discos Vaodores (CPDV), Campo Grande, 1991.]

Ministry of Aeronautics. A letter signed by **Air Force Colonel Sergio Candiota da Silva**, Assistant to the Minister of Aeronautics, to Brazilian UFO researcher Irene Granchi, dated December 19, 1988, acknowledges that the Ministry investigates UFO reports:

His Excellency recognizes the importance of the [UFO] matter, to the extent that within the Ministry of Aeronautics there exists a Bureau in charge of studying the matter, receiving, analyzing and archiving chronologically the phenomena observed in Brazilian airspace that comes to the attention of this Ministry. [Os Documentos Oficiais, ibid.]

Canada

Wilbert Smith, Senior radio engineer with the Department of Transport, headed Project Magnet, the first Canadian government UFO investigation in the 1950s. He stated in a top-secret memorandum:

The matter is the most highly classified subject in the United States Government, rating higher even than the H-bomb. Flying saucers exist. Their modus operandi is unknown but a concentrated effort is being made by a small group headed by Doctor Vannevar Bush. The entire matter is considered by the United States authorities to be of tremendous significance. [Department of Transport memorandum on "Geo-Magnetics," November 21, 1950.]

England

Sir Winston Churchill, when Prime Minister, asked to be thoroughly briefed:

> *What does all this stuff about flying saucers amount to? What can it mean? What is the truth?* [July 28, 1952, memo to Secretary of State for Air, Lord Cherwell; reprinted in Good, T., ibid.]

Prince Phillip, His Royal Highness, Duke of Edinburgh. Having been interested in the subject of UFOs since the early 1950s, he has stated:

> *There are many reasons to believe that they (UFOs) do exist: there is so much evidence from reliable witnesses.* [*Sunday Dispatch,* London, March 28, 1954.]

Air Chief Marshal Lord Dowding, Commander-in-Chief of RAF Fighter Command during the Battle of Britain, made the following comment to the press in 1954:

> *More than 10,000 sightings have been reported, the majority of which cannot be accounted for by any "scientific" explanation . . . I am convinced that these objects do exist and that they are not manufactured by any nation on earth. I can therefore see no alternative to accepting the theory that they come from some extraterrestrial source.* [*Sunday Dispatch,* London, July 11, 1954.]

Earl of Kimberly, former Liberal Party spokesman on aerospace, and member of the House of Lords:

> *UFOs defy worldly logic . . . The human mind cannot begin to comprehend UFO characteristics: their*

*propulsion, their sudden appearance, their disap-
pearance, their great speeds, their silence, their ma-
noeuvre, their apparent anti-gravity, their changing
shapes."* [House of Lords, *Debate on Unidentified
Flying Objects,* HANSARD (Lords), vol. 397, no. 23,
January 18, 1979.)

Lord Davies of Leek, Member of the House of Lords:

*If one human being out of tens of thousands who
allege to have seen these phenomena is telling the
truth, then there is a dire need for us to look into the
matter.* [Ibid.]

Lord Rankeillour, Member of the House of Lords:

*Many men have seen them [UFOs] and have not been
mistaken. Who are we to doubt their word? ... Only a
few weeks ago a Palermo policeman photographed
one, and four Italian Navy officers saw a 300-foot
long fiery craft rising from the sea and disappearing
into the sky ... Why should these men of law enforcement and defense lie?* [Ibid.]

Lord Hill-Norton (GCB), Chief of Defense Staff, Ministry
of Defense, Great Britain (1971–73); Chairman, Military
Committee of NATO (1974–77); Admiral of the Fleet;
Member of House of Lords. In 1987, he wrote the Foreword
to a book written by British UFO researcher Timothy Good,
Above Top Secret, in which he stated:

*The evidence that there are objects which have been
seen in our atmosphere, and even on terra firma, that
cannot be accounted for either as man-made objects
or as any physical force or effect known to our scientists seems to me to be overwhelming ... A very large
number of sightings have been vouched for by per-*

sons whose credentials seem to me unimpeachable.
It is striking that so many have been trained observ-
ers, such as police officers and airline or military pi-
lots. Their observations have in many instances . . .
been supported either by technical means such as
radar or, even more convincingly, by . . . interference
with electrical apparatus of one sort or another . . .
[Good, T., ibid.]

France

General Lionel M. Chassin, Commanding General of the
French Air Forces, and General Air Defense Coordinator,
Allied Air Forces, Central Europe (NATO). From 1964 until
his death in 1970, he was president of the French private
UFO research group GEPA.

The number of thoughtful, intelligent, educated peo-
ple in full possession of their faculties who have "seen
something" and described it, grows every day . . . We
can . . . say categorically that mysterious objects
have indeed appeared and continue to appear in the
sky that surrounds us.

He observed that some UFO sightings by different persons over a twenty-four-hour period, when plotted on a map,
revealed that the UFO appeared to travel in either a straight
line or a large circle. Concerning these patterns, he concluded:

Webs and networks . . . unmistakably suggest a sys-
tematic aerial exploration and cannot be the result of
chance. It indicates purposive and intelligent action.
[Chassin L., Foreword to the book by Michel Aime,
Flying Saucers and the Straight Line Mystery, New
York: Criterion Books, 1958.]

M. Robert Galley, French Minister of Defense, interviewed on radio by Jean-Claude Bourret, on February 21, 1974, stated:

> *I must say that if listeners could see for themselves the mass of reports coming in from the airborne gendarmerie, from the mobile gendarmerie, and from the gendarmerie charged with the job of conducting investigations, all of which reports are forwarded by us to the National Center for Space Studies, then they would see that it is all pretty disturbing.*
>
> *I believe that the attitude of spirit that we must adopt vis-a-vis this phenomena is an open one, that is to say that it doesn't consist in denying apriori, as our ancestors of previous centuries did deny many things that seem nowadays perfectly elementary.* [Bourret, Jean-Claude, *La nouvelle vague des soucoups volantes,* Paris: editions france-empire, 1975.]

French Space Agency. Since 1977, France has maintained an office for investigating UFO reports attached to the National Center for Space Studies (CNES) in Toulouse. The original bureau **GEPAN** (Groupe d'Etudes des Phénomènes Aérospatiaux) was changed in 1988 into the current **SEPRA** (Service d'Expertise des Phénomènes de Rentrées Atmosphériques).

GEPAN/SEPRA has analyzed the best UFO reports collected by the National Gendarmerie, relying on laboratories and scientific centers throughout France for many of its investigations. After reviewing hundreds of UFO reports collected by the Gendarmerie between 1974 and 1978, nineteen experts at GEPAN completed a report of five volumes for the GEPAN Scientific Committee, supervised by **Dr. Claude Poher**, Ph.D. in astronomy, the founder and first director of GEPAN. The expert team concluded:

> *Taking into account the facts that we have gathered from the observers and from the location of their ob-*

servations, we concluded that there generally can be said to be a material phenomenon behind the observations. In 60% of the cases reported here (references cited), the description of this phenomenon is apparently one of a flying machine whose origin, modes of lifting and/or propulsion are totally outside our knowledge.

The study of the observed phenomenon seems to us, by its extraordinary characteristics, potentially able to bring to humankind knowledge and eventually techniques of considerable importance. We suggest that a deep study of this phenomenon be undertaken with a high degree of priority. [GEPAN Report to the Scientific Committee, June 1978, vol. 1, chapter 4.]

Jean-Jacques Velasco, the last head of GEPAN and current director of SEPRA at CNES Headquarters in Toulouse. In an interview with the French magazine *Phénomèna*, Velasco stated that SEPRA's primary task was tracking "satellite re-entries, which are more and more numerous, and secondly, to continue the activities of GEPAN, stopped in 1988." He further stated:

There are cases which remain unexplained . . . Let's say simply that the events which were registered and measured, particularly at Trans-en-Provence, but also in the case of l'Amarante [a CE-II on Oct. 21, 1982] and two others, allow us to suppose that there are phenomena which escape our understanding completely. I must say that this permits us to suppose that there is an intelligence behind the phenomena. But I believe it would be largely speculation to go beyond this point. [Petrakis, Perry, and Marhic, Renaud, "Le SEPRA, côte coulisses," Phénomèna, no. 4, July–August 1991.]

Hungary

George Keleti, Minister of Defense:

> I believe that we are not alone in the universe and
> other galaxies are also carrying life on the planets. I
> never saw any alien green men here on the Earth. Yes,
> I was a columnist [in Budapest's Ufomagazin] and I
> published UFO cases that were observed and regis-
> tered within the Hungarian armed forces. I never
> stated that we are preparing any kind of action
> against UFO forces, I only pointed out to the public
> that, as a civilization, we would be unable to defend
> ourselves here on the Earth . . . Around Szolnok many
> UFO reports have been received from the Ministry of
> Defense, which obviously and logically means that
> they [UFOs] know very well where they have to land
> and what they have to do. It is remarkable indeed that
> the Hungarian newspapers, in general newspapers
> everywhere, reject the reports of the authorities.
> [Lenart, Attila, "Ask a Question to the Minister of
> Defense, George Keleti, are you afraid of a UFO in-
> vasion?" *Nepszava,* Budapest, August 18, 1994.]

Indonesia

Air Marshall Roesmin Nurjadin, Commander-in-Chief of
the Indonesian Air Force, stated in 1967:

> UFOs sighted in Indonesia are identical with those
> sighted in other countries. Sometimes they pose a prob-
> lem for our air defence and once we were obliged to
> open fire on them. [Letter to Yusuke J. Matsumura,
> May 5, 1967; reprinted in Good, T., ibid.]

Air Commander J. Salutun, Member of Parliament and Secretary of the National Aerospace Council of the Republic of Indonesia, stated in 1974:

> *I am convinced that we must study the UFO problem seriously for reasons of sociology, technology and security . . .* [Letter published in *UFO News*, vol. 6, no. 1, 1974, CBA International, Yokohama, Japan.]

Japan

General Kanshi Ishikawa, Chief of Staff of Japan's Air Self-Defense Force; Commander of the 2nd Air Wing, Chitose Air Base. In 1967, he stated:

> *Much evidence tells us UFOs have been tracked by radar; so, UFOs are real and they may come from outer space . . . UFO photographs and various materials show scientifically that there are more advanced people piloting the saucers and motherships.* [1967 interview published in *UFO News*, ibid.]

Toshiki Kaifu, Prime Minister, gave a brief interview to students of Waseda University in Tokyo in November 1989. The question of whether Japan had an official UFO organization was discussed:

> *Japan does not have such organizations at a government level . . . If young people display a serious interest in similar phenomena, we should perhaps think of forming a UFO-data collecting group under the auspices of the Ministry of Education.* [Ovsyannikov, Vladimir, "Toshiki Kaifu: 'I want to believe in my dream,' " *New Times,* Moscow, April 16–22, 1991.]

A year later, concerning an upcoming Symposium on Space and UFOs, he stated:

> *First of all, I told a magazine this past January that, as an underdeveloped country with regards to the UFO problem, Japan had to take into account what should be done about the UFO question, and that we had to spend more time on these matters. In addition, I said that someone had to solve the UFO problem with far reaching vision at the same time. Secondly, I believe it is a reasonable time to take the UFO problem seriously as a reality . . . I hope that this Symposium will contribute to peace on earth from the point of view of outer space, and take the first step toward the international cooperation in the field of UFOs. From the point of view of "people" in outer space, all human beings on earth are the same people, regardless of whether they are American, Russian, Japanese, or whoever.* [Letter to Mayor Shiotani, dated June 24, 1990, endorsing a forthcoming Space & UFO Symposium in Hakui City, Ishikawa prefecture.]

Spain

General Carlos Castro Cavero, General in the Spanish Air Force, at one time Commander of Spain's Third Aerial Region. In a 1976 interview with journalist J. J. Benítez, he acknowledged that UFOs were taken quite seriously by the Spanish military. He added:

> *Everything is in a process of investigation both in the United States and in Spain, as well as in the rest of the world . . . Look, as a General, as a military man, I have the same position as the one officially held by*

the Ministry [of Defense]. Now, from a personal position, as Carlos Castro Cavero, I believe that UFOs are spaceships or extraterrestrial craft . . . The nations of the world are currently working together in the investigation of the UFO phenomenon. There is an international exchange of data. Maybe when this group of nations acquire more precise and definite information, it will be possible to release the news to the world.

During the same interview, the General described a daytime UFO sighting at his ranch near Zaragoza:

I myself have observed one [UFO] for more than an hour . . . It was an extremely bright object, which remained stationary there for that length of time and then shot off towards Egea de los Caballeros, covering the distance of twenty kilometers in less than two seconds. No human device is capable of such a speed.

He added that the Spanish Air Ministry investigated UFO cases, including instances in which pilots had flown alongside UFOs, but when they tried to get closer, the UFOs moved away at fantastic speeds. The investigations were kept confidential at the time, but in 1992 the Spanish Air Force finally began to declassify its UFO files systematically. [Benítez, J. J., *La Gaceta del Norte,* Bilbao, Spain, June 27, 1976.]

USSR/Russia

Mikhail Gorbachev, in a speech to the International Forum, "For a Nuclear-Free World and the Survival of Humanity," at the Grand Kremlin Palace in Moscow on February 16, 1987, stated:

> *In spite of all the differences between us, we must all learn to preserve our one big family of humanity. At our meeting in Geneva, the U.S. President said that if the earth faced an invasion by extraterrestrials, the United States and the Soviet Union would join forces to repel such an invasion. I shall not dispute the hypothesis, though I think it's early yet to worry about such an intrusion.* [*Soviet Life Supplement,* May 1987.]

In reply to the question, "Does the USSR government study UFOs?" asked while visiting the Uralmash plant in Sverdlovsk on April 26, 1990, he answered:

> *I know that there are scientific organizations which study this problem.* [*Pravda,* April 27, 1990; Major General Pavel Popovich in a letter to Colman S. Von Keviczky, July 31, 1991.]

Zimbabwe

On July 22, 1985, in western Zimbabwe, a UFO was witnessed by dozens of persons on the ground and in the control tower at Bulawayo Airport, as well as by the pilots of two Hawk jets that were scrambled to pursue it. The UFO was also tracked on radar. The UFO was very bright and rounded, with a short cone above it, and evaded the Hawk jets.

Air Marshal Azim Daudpota:

> *This was no ordinary UFO. Scores of people saw it. It was no illusion, no deception, no imagination.* [*The Times,* London, August 3, 1985.]

Air Commodore David Thorne, Director General of Operations in a October 24, 1985, letter to Timothy Good, stated:

> *[Although not speaking officially], as far as my Air Staff is concerned, we believe implicitly that the unexplained UFOs are from some civilization beyond our planet.* [Good, T., ibid.]

ASTRONAUTS

United States

Gordon Cooper, Astronaut (Mercury-Atlas 9, May 15, 1963; Gemini 5, August 21, 1965), Col. USAF (Ret); letter to Granada's Ambassador Griffith at the United Nations, November 9, 1978:

> I wanted to convey to you my views on our extra-terrestrial visitors popularly referred to as "UFOs," and suggest what might be done to properly deal with them.
>
> I believe that these extra-terrestrial vehicles and their crews are visiting this planet from other planets, which obviously are a little more technically advanced than we are here on earth. I feel that we need to have a top level, coordinated program to scientifically collect and analyze data from all over the earth concerning any type of encounter, and to determine how best to interface with these visitors in a friendly fashion. We may first have to show them that we have learned to resolve our problems by peaceful means, rather than warfare, before we are accepted as fully qualified universal team members. This acceptance would have tremendous possibilities of advancing our world in all areas. Certainly then it would seem that the UN has a vested interest in handling this subject properly and expeditiously.
>
> I should point out that I am not an experienced UFO professional researcher. I have not yet had the privilege of flying a UFO, nor of meeting the crew of one. I do feel that I am somewhat qualified to discuss them since I have been into the fringes of the vast

areas in which they travel. Also, I did have occasion in 1951 to have two days of observation of many flights of them, of different sizes, flying in fighter formation, generally from east to west over Europe. They were at a higher altitude than we could reach with our jet fighters of that time.

I would also like to point out that most astronauts are very reluctant to even discuss UFOs due to the great numbers of people who have indiscriminately sold fake stories and forged documents abusing their names and reputations without hesitation. Those few astronauts who have continued to have participation in the UFO field have had to do so very cautiously. There are several of us who do believe in UFOs and who have had occasion to see a UFO on the ground, or from an airplane. There was only one occasion from space which may have been a UFO.

If the UN agrees to pursue this project, and to lend their credibility to it, perhaps many more well qualified people will agree to step forth and provide help and information. [Good, T., ibid.]

On several occasions, he described his own sighting in Germany in 1951:

Several days in a row we sighted groups of metallic, saucer-shaped vehicles at great altitudes over the base, and we tried to get close to them, but they were able to change direction faster than our fighters. I do believe UFOs exist and that the truly unexplained ones are from some other technologically advanced civilization. From my association with aircraft and spacecraft, I think I have a pretty good idea of what everyone on this planet has and their performance capabilities, and I'm sure some of the UFOs at least are not from anywhere on Earth. [*Omni*, vol. 2, no. 6, March 1980.]

Donald (Deke) Slayton, Mercury 7 astronaut, stated in an interview that he had seen a UFO in 1951:

> *I was testing a P-51 fighter in Minneapolis when I spotted this object. I was at about 10,000 feet on a nice, bright, sunny afternoon. I thought the object was a kite, then realized that no kite is gonna* [sic] *fly that high. As I got closer, it looked like a weather balloon, gray and about three feet in diameter. But as soon as I got behind the darn thing, it didn't look like a balloon anymore. It looked like a saucer, a disc. About that same time, I realized that it was suddenly going away from me—and there I was, running at about 300 miles an hour. I tracked it for a little while, and then all of sudden the damn thing just took off. It pulled about a 45-degree climbing turn and accelerated and just flat disappeared. A couple of days later, I was having a beer with my commanding officer, and I thought, "What the hell, I'd better mention something to him about it." I did, and he told me to get on down to intelligence and give them a report. I did, and I never heard anything more on it.* [National Enquirer, October 23, 1979.]

Edgar Mitchell, Apollo 14 lunar module pilot and founder of the Institute of Noetic Sciences:

> *I've changed my position in the last two or three years—the last two years to be precise—to suggest that the evidence is strong enough that we really need to have serious open discussion and release of information that it is quite clear the government and other governments do hold, and that this become a part of our official knowledge . . . Now, whether it's true or not, it deserves to be handled with a serious investigation. There is too much smoke here not to be fire,*

*and so I personally in the last couple of years have
come out—I don't know the answers, but I've come
out—and I say, this has gone far enough. If it's real,
let's get it out in the open; let's break the deadlock
that bureaucracy has on this. There is enough evidence pointing in the direction that clearly there is information being withheld. How far we can go with it,
I don't know.* [Except from his lecture "Science and
the Inner Experience" sponsored by the friends of the
Institute of Noetic Sciences, New York City, December 4, 1991.]

Al Worden, Apollo 15 astronaut who later became a poet. In
a lengthy interview in a documentary produced for the twentieth anniversary of the landing on the Moon, Worden discussed his views that Earth was probably visited in the past
by extraterrestrial explorers. He began by commenting on
the well-known "UFO interpretation" of the vision of the
prophet Ezekiel in the Bible:

*And a literal translation describes very clearly a
spacecraft with the ability to land vertically and take-
off vertically, and it was an object that looked very
much like the Lunar Module that we used on the
Moon; and if it's going to land vertically and take-off
vertically, it had to come from some place and go
back some place.*

*In my mind the universe has to be cyclic, so that in
one galaxy if there is a planet maybe that has arrived
at the point of becoming unlivable, you will find in
another part of a different galaxy a planet that has
just formed which is perfect for habitation. I see some
kind of intelligent being, like us, skipping around the
universe from planet to planet as, let's say, the South
Pacific Indians do on the islands, where they skip
from island to island. When the first island blows up*

due to a volcano, they will have their progeny on all these other islands and they will be able to continue the species. I think that's what the [alien] space program is all about.

I think we may be a combination of creatures that were living here on Earth some time in the past, and having a visitation, if you will, by creatures from somewhere else in the universe, and those two species getting together and having progeny. I am not at all convinced that we are not the result of that particular union some many thousands of years ago. If that is the case in fact, a very small group of explorers could land on a planet and create successors to themselves that would eventually take up the pursuit of, let's say, inhabiting the rest of the universe. [Excerpts from his interview in the documentary "The Other Side of the Moon," produced by Michael G. Lemle, and broadcast by PBS in July 1989.]

Eugene Cernan, Apollo 17 Commander:

I'm one of those guys who has never seen a UFO. But I've been asked, and I've said publicly I thought they were somebody else; some other civilization . . . [Chriss, Nicholas, "Cernan Says Other Earths Exist," *Los Angeles Times,* January 6, 1973.]

Story Musgrave, Space shuttle astronaut who flew on the repair mission of the Hubble Space Telescope:

I try to communicate with the life that's out there. I'm serious. It is not that far out. When I'm circling around out there, I try in whatever ways I can to get them to come down here and get me." [*The Houston Post,* December 1, 1993.]

USSR/Russia

Yevegni Khrunov, Soyuz-5 spacecraft pilot in 1969:

Is the presence of extraterrestrial civilizations con-
ceivable? Of course. Before the uniqueness of the earth
is demonstrated, this assumption should be taken as
quite legitimate. As regards UFOs, their presence
cannot be denied: thousands of people have seen
them. It may be that their source is optical effects, but
some of their properties, for instance, their ability to
change course by 90 degrees at great speed, simply
stagger the imagination. [*Sputnik,* "UFOs Through
the Eyes of Cosmonauts," December 1980.]

Vladimir Kovalyonok, Major-General of Aviation, stated:

On May 5, 1981, we were in orbit [in the Salyut-6
space station]. I saw an object that didn't resemble
any cosmic objects I'm familiar with. It was a round
object which resembled a melon, round and a little bit
elongated. In front of this object was something that
resembled a gyrating depressed cone. I can draw it,
it's difficult to describe. The object resembles a bar-
bell. I saw it becoming transparent and like with a
"body" inside. At the other end I saw something like
gas discharging, like a reactive object. Then some-
thing happened that is very difficult for me to de-
scribe from the point of view of physics. Last year in
the magazine Nature I read about a physicist . . . we
tried together to explain this phenomenon and we
decided it was a "plasmaform." I have to recognize
that it did not have an artificial origin. It was not arti-
ficial because an artificial object couldn't attain this
form. I don't know of anything that can make this
movement . . . tightening, then expanding, pulsating.

Then as I was observing, something happened, two explosions. One explosion, and then 0.5 seconds later, the second part exploded. I called my colleague Viktor [Savinykh], but he didn't arrive in time to see anything.

What are the particulars? First conclusion: the object moved in a suborbital path, otherwise I wouldn't have been able to see it. There were two clouds, like smoke, that formed a barbell. It came near me and I watched it. Then we entered in to the shade for two or three minutes after this happened. When we came out of the shade we didn't see anything. But during a certain time, we and the craft were moving together. [Videotaped interview with Giorgio Bongiovanni in the village of Kosnikov, near Moscow, 1993. Videotape courtesy of Michael Hesemann.]

Major-General Pavel Popovich, pioneer cosmonaut, "Hero of the Soviet Union," and the President of All-Union Ufology Association of the Commonwealth of Independent States:

Today it can be stated with a high degree of confidence that observed manifestations of UFOs are no longer confined to the modern picture of the world, or the simple refutation of the orthodox natural science paradigm. The historical evidence of the phenomenon, the singularity of its newly gained kinematic, energetic, and psychophysical features allows us to hypothesize that ever since mankind has been coexisting with this extraordinary substance, it has manifested a high level of intelligence and technology. The UFO sightings have become the constant component of human activity and require a serious global study. In order to realize the position of man on earth and in the universe, ufology, the scientific study of the UFO phenomenon, should take place in the midst of other sciences dealing with man and the world . . .

The influence the UFO has on people, as well as the effects it produces, should become the items of special research. The UFO's interaction with the environment, the behavior that it motivates, and its genesis, also present interesting areas for concentrated study. Today, many specialists have come to the opinion that [UFO] phenomenon research should be taken up along with understanding and comprehension of other unexplained phenomena . . . The development of new approaches for the identification and study of energy and information processes will allow for an enthusiastic move toward the comprehension of the phenomenon. The results of these studies should aid the survival of the people on earth . . .

It's necessary to carry out the popular ufological enlightenment, since the probability for a meeting of a person with a UFO exists, and this person should be ready for this event. Precautionary measures are especially important. It's necessary to tell the truth, which has been distorted previously by the politically engaged sciences and most recently by ufological dilettantes. The main purpose of the primary local groups, that of controlling the ufological situation, mustn't be forgotten. The ufologists should know all the UFO's landing places and contacts in their regions. They should have relations with the local authorities, and in particular, with the police, the civil defense bodies, as well as information, scientific, and medical organizations. [Popovich, P., "Ufology in the Commonwealth of Independent States: Organization Problems," in the *MUFON 1992 International UFO Symposium Proceedings*.]

SCIENTISTS

United States

Dr. Clyde W. Tombaugh, American astronomer who discovered the planet Pluto. On August 20, 1949, he observed a UFO that appeared as a geometrically arranged group of six to eight rectangles of light, windowlike in appearance and yellowish green in color, which moved from northwest to southeast over Las Cruces, New Mexico. He stated:

> I doubt that the phenomenon was any terrestrial reflection, because . . . nothing of the kind has ever appeared before or since . . . I was so unprepared for such a strange sight that I was really petrified with astonishment. [*The UFO Evidence*, ibid.]

Dr. Carl Sagan, former Professor of Astronomy and Space Sciences at Cornell University, stated:

> It now seems quite clear that Earth is not the only inhabited planet. There is evidence that the bulk of the stars in the sky have planetary systems. Recent research concerning the origin of life on Earth suggests that the physical and chemical processes leading to the origin of life occur rapidly in the early history of the majority of planets. The selective value of intelligence and technical civilization is obvious, and it seems likely that a large number of planets within our Milky Way galaxy—perhaps as many as a million—are inhabited by technical civilizations in advance of our own. Interstellar space flight is far beyond our present technical capabilities, but there seems to be no fundamental physical objections to preclude, from our own

*vantage point, the possibility of its development by
other civilizations.* [Sagan, Carl, "Unidentified Flying
Objects," *The Encyclopedia Americana*, 1963.]

Dr. Frank B. Salisbury, Professor of Plant Physiology at
Utah State University:

> *I must admit that any favorable mention of the flying
> saucers by a scientist amounts to extreme heresy and
> places the one making the statement in danger of ex-
> communication by the scientific theocracy. Neverthe-
> less, in recent years I have investigated the story of the
> unidentified flying object (UFO), and I am no longer
> able to dismiss the idea lightly.* [Paper on "Exobiol-
> ogy," presented at the First Annual Rocky Mountain
> Bioengineering Symposium, held at the United States
> Air Force Academy, in May 1964. Quoted in Fuller,
> John G., *Incident at Exeter,* Putnam, 1966.]

Dr. J. Allen Hynek, Chairman of the Department of Astronomy at Northwestern University and scientific consultant for Air Force investigations of UFOs from 1948 until
1969 (Projects Sign, Grudge, and Blue Book). Over his long
career, he made numerous comments about the scientific implications of the UFO phenomenon:

> *Despite the seeming inanity of the subject, I felt that I
> would be derelict in my scientific responsibility to the
> Air Force if I did not point out that the whole UFO phe-
> nomenon might have aspects to it worthy of scientific
> attention.* [Hearings on Unidentified Flying Objects,
> Committee on Armed Services, House of Representa-
> tives, Eighty-ninth Congress, Second session, 1966.]
>   *I have begun to feel that there is a tendency in 20th
> Century science to forget that there will be a 21st Cen-
> tury science, and indeed a 30th Century science, from
> which vantage points our knowledge of the universe*

*may appear quite different than it does to us. We suffer,
perhaps, from temporal provincialism, a form of arro-
gance that has always irritated posterity.* [Hynek, J.
Allen, letter to *Science* magazine, August 1, 1966.]

*When one gets reports from scientists, engineers
and technicians whose credibility by all common
standards is high and whose moral caliber seems to
preclude a hoax, one can do no less than hear them
out, in all seriousness.* [Hynek, J. Allen, "The UFO
Gap," *Playboy,* vol. 14, no. 12, December 1967.]

*There exists a phenomenon . . . that is worthy of
systematic rigorous study . . . The body of data point to
an aspect or domain of the natural world not yet explored by science . . . When the long awaited solution
to the UFO problem comes, I believe that it will prove
to be not merely the next small step in the march of
science but a mighty and totally unexpected quantum
jump.* [Hynek, J. Allen, *The UFO Experience: A Scientific Inquiry,* Chicago: Regnery Co., 1972.]

Dr. Leo Sprinkle, Professor of Psychology at the University
of Wyoming, had his first UFO sighting in 1951, when he
and a friend saw "something in the sky, round and metallic
looking." In 1956, he had a second sighting while driving
with his wife near Boulder, Colorado:

*We watched it for quite a few minutes. We could see it
was larger than the headlights of the cars below. And
we could see it was not attached to anything. And
there was no sound. I became frightened actually, because it wasn't anything I could understand . . . from
a personal viewpoint, I am pretty well convinced that
we are being surveyed.* ["Flying Saucers," Special
Issue of *Look* magazine, 1967.]

Dr. James E. McDonald, Senior Physicist at the Institute of
Atmospheric Physics at the University of Arizona, testified

at the UFO hearings convened by the House Committee on Science and Astronautics in 1968:

> *The type of UFO reports that are most intriguing are close-range sightings of machine-like objects of unconventional nature and unconventional performance characteristics, seen at low altitudes, and sometimes even on the ground. The general public is entirely unaware of the large number of such reports that are coming from credible witnesses . . . When one starts searching for such cases, their number are quite astonishing. Also, such sightings appear to be occurring all over the globe.* ["Symposium on Unidentified Flying Objects," Hearings before the Committee on Science and Astronautics, U.S. House of Representatives, July 29, 1968.]

Dr. Robert M. L. Baker, Jr., President of West Coast University; author of two astrodynamics textbooks; head of Lockheed's Astrodynamics Research Center (1961–64); member of the faculty of Astronomy and Engineering at UCLA (1959–71).

He has specialized in the study of motion pictures of UFOs and anomalistic radar images and has concluded that two of the most famous UFO motion pictures, taken in the 1950s, cannot be explained in terms of conventional phenomena.

In 1968, he made the following statement concerning the one U.S. radar system in operation at that time that, to his knowledge, exhibited sufficient continuous coverage to reveal UFOs operating above the Earth's atmosphere:

> *The system is partially classified and, hence, I cannot go into great detail . . . Since this particular sensor system has been in operation, there have been a number of anomalistic alarms. Alarms that, as of this date, have not been explained on the basis of natural phenomena interference, equipment malfunction or*

inadequacy, or man-made space objects. [1968 Congressional Hearings, ibid.]

Stanton T. Friedman, Nuclear physicist and well-known UFO researcher, responsible for the original investigation of the Roswell, New Mexico, incident. In a prepared statement submitted to the House Science and Astronautics Committee UFO Hearings in 1968, he posed and answered a series of key questions about the UFO phenomenon:

1. To what conclusions have you come with regard to UFOs?

I have concluded that the earth is being visited by intelligently controlled vehicles whose origin is extraterrestrial. This doesn't mean I know where they come from, why they are here, or how they operate.

2. What basis do you have for these conclusions?

Eyewitness and photographic and radar reports from all over the earth by competent witnesses of definite objects whose characteristics such as maneuverability, high speed, and hovering, along with definite shape, texture, and surface features, rule out terrestrial explanations.

3. Were there any differences between the unknowns and the knowns?

A "chi square" statistical analysis was performed comparing the unknowns in this study to all the knowns. It was shown that the probability that the unknowns came from the same population of sighting reports as the knowns, was less than 1%. This was based on apparent color, velocity, etc. . . . Maneuverability, one of the most distinguished characteristics

of UFOs, was not included in this statistical analysis.
[1968 Congressional Hearings, ibid.]

Dr. Margaret Mead, World-renowned anthropologist, stated:

There are unidentified flying objects. That is, there are a hard core of cases—perhaps 20 to 30 percent in different studies—for which there is no explanation . . . We can only imagine what purpose lies behind the activities of these quiet, harmlessly cruising objects that time and again approach the earth. The most likely explanation, it seems to me, is that they are simply watching what we are up to. [Mead, Margaret, "UFOs—Visitors from Outer Space?" *Redbook,* vol. 143, September 1974.]

American Institute of Aeronautics and Astronautics UFO Subcommittee. The AIAA established a subcommittee in 1967 to look into the UFO question. The UFO Subcommittee issued several reports and statements, including in-depth studies of two UFO incidents. The UFO Subcommittee stated that its "most important conclusion" was that government agencies consider funding UFO research:

From a scientific and engineering standpoint, it is unacceptable to simply ignore substantial numbers of unexplained observations . . . the only promising approach is a continuing moderate-level effort with emphasis on improved data collection by objective means . . . involving available remote sensing capabilities and certain software changes. [Story, Ronald D., *The Encyclopedia of UFOs,* New York: Doubleday, 1980.]

The Subcommittee of the American Institute of Aeronautics and Astronautics criticized the conclusion of the *Condon Report* as the personal views of Dr. Condon, and added:

*The opposite conclusion could have been drawn from
The Condon Report's content, namely, that a phenomenon with such a high ratio of unexplained cases
(about 30 percent) should arouse sufficient scientific
curiosity to continue its study.* [Ibid.]

Dr. Richard Haines, Psychologist specializing in pilot and
astronaut "human factors" research for the Ames NASA Research Center in California, from which he retired in 1988 as
Chief of the Space Human Factors Office. He stated:

*We're not dealing with mental projections or halluci-
nations on the part of the witness but with a real
physical phenomenon.* [Haines, Dr. Richard, *Observing UFO's,* Chicago: Nelson-Hall, 1980.]

A principal focus of his UFO research concerns aircraft
cases:

*Reports of anomalous aerial objects (AAO) appearing
in the atmosphere continue to be made by pilots of almost every airline and air force of the world in addition to private and experimental test pilots. This paper
presents a review of 56 reports of AAO in which electromagnetic effects (E-M) take place on-board the aircraft when the phenomenon is located nearby but not
before it appeared or after it had departed.*

*Reported E-M effects included radio interference
or total failure, radar contact with and without si-
multaneous visual contact, magnetic and/or gyro-
compass deviations, automatic direction finder failure
or interference, engine stopping or interruption, dim-
ming cabin lights, transponder failure, and military
aircraft weapon system failure.* [Haines, Dr. Richard,
"Fifty-Six Aircraft Pilot Sightings Involving Electromagnetic Effects," *MUFON 1992 International UFO
Symposium Proceedings.*]

Dr. Peter A. Sturrock, Professor of Space Science and Astrophysics and Deputy Director of the Center for Space Sciences and Astrophysics at Stanford University; Director of the Skylab Workshop on Solar Flares in 1977. He stated:

The definitive resolution of the UFO enigma will not come about unless and until the problem is subjected to open and extensive scientific study by the normal procedures of established science. This requires a change in attitude primarily on the part of scientists and administrators in universities. [Sturrock, Peter A., *Report on a Survey of the American Astronomical Society concerning the UFO Phenomenon,* Stanford University Report SUIPR 68IR, 1977.]

Although . . . the scientific community has tended to minimize the significance of the UFO phenomenon, certain individual scientists have argued that the phenomenon is both real and significant. Such views have been presented in the Hearings of the House Committee on Science and Astronautics [and elsewhere]. It is also notable that one major national scientific society, the American Institute of Aeronautics and Astronautics, set up a subcommittee in 1967 to "gain a fresh and objective perspective on the UFO phenomenon."

In their public statements (but not necessarily in their private statements), scientists express a generally negative attitude towards the UFO problem, and it is interesting to try to understand this attitude. Most scientists have never had the occasion to confront evidence concerning the UFO phenomenon. To a scientist, the main source of hard information (other than his own experiments' observations) is provided by the scientific journals. With rare exceptions, scientific journals do not publish reports of UFO observations. The decision not to publish is made by the editor acting on the advice of reviewers. This process is self-reinforcing: the apparent lack of data confirms the view that there is nothing to the UFO phenomenon, and this view works against the presentation of relevant data. [Sturrock, Peter A., "An Analysis of the Condon Report on the Colorado UFO Project," *Journal of Scientific Exploration,* vol. 1, no. 1, 1987.]

Dr. Jacques Vallee, Astrophysicist, computer scientist, and world-renowned researcher and author on UFOs and paranormal phenomena. He worked closely with Dr. J. Allen Hynek. Commenting on the need for science "to search beyond the superficial appearances of reality":

Skeptics, who flatly deny the existence of any unexplained phenomenon in the name of "rationalism," are among the primary contributors to the rejection of science by the public. People are not stupid and they know very well when they have seen something out of the ordinary. When a so-called expert tells them the object must have been the moon or a mirage, he is really teaching the public that science is impotent or unwilling to pursue the study of the unknown. [Vallee, J., *Confrontations,* New York: Ballantine Books, 1990.]

Vallee reveals from his diaries how the government has deliberately misled the scientific world, the media, and the public regarding their information on UFOs and paranormal research:

It is unusual for scientists to keep diaries and even more unusual for them to make them public . . . I have followed this rule of silence for the last thirty years, but I have finally decided that I had no right to keep them private anymore . . . They provide a primary source about a crucial fact in the recent historical record: the appearance of new classes of phenomena that highlighted the reality of the paranormal. These

*phenomena were deliberately denied or distorted by
those in authority within the government and the
military. Science never had fair and complete access
to the most important files.*

*The thirteen years covered here, from 1957 to
1969, saw some of the most exciting events in technological history . . . Behind the grand parade of the
visible breakthroughs in science, however, more private mysteries were also taking place: . . . all over the
world people had begun to observe what they described as controlled devices in the sky. They were
shaped like saucers or spheres. They seemed to vio-
late every known principle in our physics.*

*Governments took notice, organizing task forces,
encouraging secret briefings and study groups, funding classified research and all the time denying before the public that any of the phenomena might be
real . . . The major revelation of these Diaries may be
the demonstration of how the scientific community
was misled by the government, how the best data
were kept hidden, and how the public record was
shamelessly manipulated.* [Vallee, J., *Forbidden Science*, Berkeley: North Atlantic Books, 1992.]

Dr. John E. Mack, Professor of Psychiatry at the Cam-
bridge Hospital, Harvard Medical School, and founding
director of the Center for Psychology and Social Change. A
1977 Pulitzer Prize–winner for his biography of Lawrence
of Arabia, Dr. Mack has studied the subject of UFO abductions in recent years:

*I will stress once again that we do not know the
source from which the UFOs or the alien beings come
(whether or not, for example, they originate in the
physical universe as modern astrophysics has described it). But they manifest in the physical world*

*and bring about definable consequences in that do-
main.* [Mack, J., *Abduction—Human Encounters With
Aliens,* New York: Scribners, 1994.]

Belgium

Dr. Auguste Meessen, Professor of Physics at the Catholic
University in Louvain and one of the scientific consultants
for the Belgian Society for the Study of Space Phenomena
(SOBEPS). In an interview with French journalist Marie-
Therese de Brosses, Professor Meessen discussed the recent
UFO wave in Belgium:

> *There are too many independent eyewitness reports
> to ignore. Too many of the reports describe coherent
> physical effects, and there is an agreement among the
> accounts concerning what was observed . . . But of
> course there are also physical effects. The Air Force
> report [of the F-16 jet scramble incident on the night
> of March 30–31, 1990] allows us to approach the
> problem in a rational and scientific way. The simplest
> hypothesis is that the reports are caused by extrater-
> restrial visitors, but that hypothesis carries with it
> other problems. We are not in a rush to form a con-
> clusion, but continue to study the mystery.* [Brosses,
> M.-T. de, "F-16 Radar Tracks UFO," *Paris Match,*
> July 5, 1990. English version in the *MUFON UFO
> Journal,* no. 268, August 1990.]

China

Chinese Academy of Social Sciences. One of the branches
of the Chinese Academy of Social Sciences is the **China**

UFO Research Organization (CURO). In 1985, CURO had 20,000 members, and two publications, the *Journal of UFO Research* and *Space Exploration*. The journal's first issue in 1981 included an article by **Comrade Bang Wen-Gwang** of the Chinese Academy of Sciences' Beijing Astronomical Research Society. The article stated in part:

> *In this field [Ufology], prejudice will take you farther from the truth than ignorance . . . But with a topic such as UFOs, where does the scientific method begin? And where does it end? This grand endeavor would consist of the serious recording of the enormous available data and the use of all scientific procedures for the purpose of analysis . . . China is so vast, and UFOs are certainly being witnessed again and again all throughout China, and China most definitely will evolve her own indigenous school of UFO researchers. This is our sincerest and deepest hope.*
> [Wen-Gwang, B., "The Aspirations & Hopes of the Chinese UFO Investigator," *The Journal of UFO Research,* no. 1, People's Republic of China, 1981.]

UFO Scientific Conference in Darlian. In 1985, the government newspaper, *China Daily,* reported that a UFO Scientific Conference was held in Darlian, with some forty papers presented on various aspects of UFO research. **Professor Liang Renglin** of Guangzhou Jinan University, Chairman of CURO, stated in the Darlian Conference that more than six hundred UFO reports had been made in China during the past five years. The article concluded:

> *UFOs are an unresolved mystery with profound influence in the world.* ["UFO Conference Held in Darlian," *China Daily,* August 27, 1985; quoted in Good, T., ibid.]

France

Dr. Pierre Guérin, Senior researcher at the French National Council for Scientific Research (CNRS), has written extensively about the need for scientific research in the UFO field. He concluded a summary of the UFO evidence published in *Sciences & Avenir* in 1972 with the following words:

> At the very least, it is already possible to show scientifically the evidence for physico-chemical modifications affecting sometimes the ground of alleged landing sites, as well as the effects produced on the vegetation. Such research has already begun and doesn't necessarily require large sums.
> The UFO problem in its totality, nevertheless, cannot be really understood unless our science someday is able to propose physical models that take into account the observed phenomena. We are not able to know if this will ever occur, and in any event, we are still very far from that stage. [Guérin, P., "Le Dossier des Objets Volants Non Identifiés," *Sciences & Avenir,* no. 307, Paris, September 1972.]

Dr. Claude Poher, Expert on aeronautics, astronomy, and astronautics, engineer at the French Space Agency (CNES) for thirty years, specializing in rocket propulsion and nuclear space energy; former chairman of many working groups in the International Astronautical Federation; founder of GEPAN in May 1977 and its first Director until 1979.

Before creating GEPAN, he studied the UFO phenomenon for many years and had access to French military and police UFO files, including classified reports. In a report on UFOs for French officials, he wrote:

> The phenomenon seems to be real . . . The general coherence of sighting reports worldwide should not leave

researchers indifferent. One does not conceive objective arguments to justify an attitude that would avoid at all cost these observations . . . The risk is, at worst, to confirm the existence of unknown vehicles appearing erratically into our atmosphere—a hypothesis that seems to explain nearly all reported aspects of the phenomenon and could be linked to the current (1970) exobiology branch of space research. [1971 Statistical Study prepared for the CNES and French officials.]

He further comments on the science and technology implied by the eyewitness descriptions of the phenomenon:

Given the volume of the objects described in the observations . . . I can affirm that our futuristic space generators are far from being able to produce the amount of energy seen by the UFO witnesses. The light power seen is probably the tip of the iceberg, because no thermodynamic system can produce energy without dissipating a part of it. The megawatts of observed light are most likely the energy "leak" from the energy conversion system used by the flying object, which means that the useful energy produced is much greater than what is seen.

The knowledge of such an energy production method is crucial for the future of mankind. The UFO observation reports tell us that ambitious, entirely new, solutions <u>are possible</u> [underlined in the original]. This is very important. [Letter to Marie Galbraith, November 26, 1995.]

Germany

Professor Hermann Oberth, German rocket expert considered (with Robert Goddard and Konstantin Tsiolkovsky) one of the three fathers of the space age. In 1955, Dr. Werner von Braun invited him to the United States, where he worked on rockets with the Army Ballistic Missile Agency and later NASA. Oberth's active interest in UFOs began in 1954:

> It is my thesis that flying saucers are real and that they are space ships from another solar system. I think that they possibly are manned by intelligent observers who are members of a race that may have been investigating our earth for centuries. I think that they possibly have been sent out to conduct systematic, long-range investigations, first of men, animals, vegetation, and more recently of atomic centers, armaments and centers of armament production. [Oberth H., "Flying Saucers Come From A Distant World," *The American Weekly,* October 24, 1954.]

As a rocket scientist and space pioneer, Professor Oberth paid close attention to the propulsion aspect of UFO research:

> They are flying by means of artificial fields of gravity . . . They produce high-tension electric charges in order to push the air out of their paths, so it does not start glowing, and strong magnetic fields to influence the ionized air at higher altitudes. First, this would explain their luminosity . . . Secondly, it would explain the noiselessness of UFO flight . . . Finally, this assumption also explains the strong electrical and magnetic effects sometimes, though not always, observed in the vicinity of UFOs. ["Dr. Hermann Oberth discusses UFOs," *Fate,* May 1962.]
>
> It is my conclusion that UFOs do exist, are very real, and are spaceships from another or more than one solar system. They are possibly manned by intelligent observers who are members of a race carrying

*out long-range scientific investigations of our earth
for centuries. [UFO News, 1974.]*

Greece

Dr. Paul Santorini, Greek physicist and engineer credited
with developing the proximity fuse for the Hiroshima atomic
bomb, two patents for the guidance system used in the U.S.
Nike missiles, and a centimetric radar system.

He stated that he believes UFOs are under intelligent
control. In 1947, he investigated a series of UFO reports
over Greece that were initially thought to be Soviet missiles:

*We soon established that they were not missiles. But,
before we could do any more, the Army, after confer-
ring with foreign officials, ordered the investigation
stopped. Foreign scientists flew to Greece for secret
talks with me . . . A world blanket of secrecy surrounded the UFO question because the authorities were unwilling to admit the existence of a force
against which we had no possibility of defense.
[Fowler, R., UFOs: Interplanetary Visitors, New York:
Bantam Books, 1974.]*

Switzerland

Dr. Carl Gustav Jung, Pioneer of psychiatry, stated in 1954:

*A purely psychological explanation is ruled out . . .
the discs show signs of intelligent guidance, by quasi-
human pilots . . . the authorities in possession of important information should not hesitate to enlighten*

the public as soon and as completely as possible. ["Dr. Carl Jung on Unidentified Flying Objects," *Flying Saucer Review,* vol. 1, no. 2, 1955.]

Unfortunately, however, there are good reasons why the UFOs cannot be disposed of in this simple manner. It remains an established fact, supported by numerous observations, that UFOs have not only been seen visually but have also been picked up on the radar screen and have left traces on the photographic plate. It boils down to nothing less than this: that either psychic projections throw back a radar echo, or else the appearance of real objects affords an opportunity for mythological projections. ["A Fresh Look at Flying Saucers," *Time,* August 4, 1967.]

USSR/Russia

Dr. Felix Y. Zigel, Professor of Mathematics and Astronomy at the Moscow Aviation Institute, known as the father of Russian Ufology. In a November 10, 1967, broadcast on Moscow Central Television, with Soviet Air Force General Porfiri Stolyarov, Zigel stated:

Unidentified flying objects are a very serious subject which we must study fully. We appeal to all viewers to send us details of strange flying craft seen over the territories of the Soviet Union. This is a serious challenge to science and we need the help of all Soviet citizens. [Good, T., ibid.]

Observations show that UFOs behave "sensibly." In a group formation flight, they maintain a pattern. They are most often spotted over airfields, atomic stations and other very new engineering installations. On encountering aircraft, they always maneuver so as to avoid direct contact. A considerable list of these *seemingly intelligent actions gives the impression that UFOs are investigating, perhaps even reconnoitering . . . The important thing now is for us to discard any preconceived notions about UFOs and to organize on a global scale a calm, sensation-free and strictly scientific study of this strange phenomenon. The subject and aims of the investigation are so serious that they justify all efforts. It goes without saying that international cooperation is vital.* [Zigel, F., "Unidentified Flying Objects," *Soviet Life,* no. 2 (137), February 1968.]

In an interview with Henri Gris in 1981, he stated:

We have seen these UFOs over the USSR; craft of every possible shape: small, big, flattened, spherical. They are able to remain stationary in the atmosphere or to shoot along at 100,000 kilometers per hour . . . They are also able to affect our power resources, halting our electricity generating plants, our radio stations, and our engines, without however leaving any permanent damage. So refined a technology can only be the fruit of an intelligence that is indeed far superior to man. [*Gente,* July 31, 1981, and August 7, 1981.]

Institute of Space Research of the Soviet Academy of Sciences published, in 1979, a 74-page statistical analysis of over 250 UFO cases reported in the Soviet Union. After stating that hallucinations, errors, and conventional explanations (aircraft, satellites, etc.) could not account for many of the reports, the study concluded:

Obviously, the question of the nature of the anomalous phenomena still should be considered open. To obtain more definite conclusions, more reliable data must be available. Reports on observations of anomalous phenomena have to be well documented. The

*production of such reports must be organized through
the existing network of meteorological, geophysical,
and astronomical observation stations, as well as
through other official channels . . . In our opinion, the
Soviet and foreign data accumulated so far justifies
setting such studies.* [Gindilis, L. M., Men'kov, D.A.
and Petrovskaya, I.G., "Observations of Anomalous
Atmospheric Phenomena in the USSR: Statistical
Analysis," USSR Academy of Sciences Institute of
Space Research, Report PR 473, Moscow, 1979. The
English translation of the full report, in "NASA Tech-
nical Memorandum No. 75665," was reprinted by the
Center for UFO Studies in June 1980.]

USSR Scientific Commissions. The Soviet press was informed in the mid-1980s that the All-Union Council of Scientific and Technical Societies (now the Council of Scientific
and Engineering Societies) had set up a nongovernmental
Commission on Paranormal Events, headed by V. S. Troitsky,
a Corresponding Member of the USSR Academy of Sciences.
A glimpse at the activities of the Commission was published
in 1989 by members **A. Petukhov** and **T. Faminskaya**:

*Of special value are the archives set up by the Com-
mission. They contain over 13 thousand reports connected with PEs [Paranormal Events] and with
UFOs in particular . . . UFOs have been seen to
hover over ground objects, to chase or fly side by side
with airplanes and cars, to follow geometrically
regular trajectories, and to send out ordered flashes
of light. In other words, such "paranormals" behave,
from the viewpoint of human beings, quite often
showing capabilities yet beyond the reach of the
machines built on the Earth.* [Faminskaya, T. and
Petukhov, A., "At 4.10 Hours and After," *Almanac
Phenomenon 1989,* Moscow Mir, 1989.]

APPENDICES

U.S. GOVERNMENT UFO
PROJECTS AND STUDIES

The involvement of the U.S. government in the UFO mystery dates back to the latter part of World War II, when foo fighters (luminous orbs at night, shiny and reflective in daylight) puzzled Allied airmen by approaching and pacing their aircraft during missions, then suddenly darting away. The objects, though assumed by debriefing intelligence officers to be enemy weapons or observation devices, never posed a threat. Sightings of them were recorded in military-unit records, but it is not clear that they were ever systematically studied.

When the first major, well-recorded UFO sighting wave began in July 1947 in the Pacific Northwest, the reports stirred memories of foo fighters among World War II veterans. Once again shiny, maneuverable unidentified objects were reported to be pacing aircraft and widely seen by ground observers as well. When thousands of citizens reported daylight sightings of disc- and oval-shaped, apparently metallic objects coursing through the skies, the Army and the spin-off Air Force (formerly Army Air Corps) initiated urgent studies.

At first it was feared that the Soviet Union, despite its bedraggled state, had somehow made a major aeronautical breakthrough—perhaps with the assistance of captured German engineers. At the onset of the Cold War, this posed a threat to U.S. and Allied interests. The initial readings quickly ruled out a Soviet origin but left an important mystery. These early intelligence findings remained totally secret for many years.

The top-level evaluations produced such conclusions as: "This 'flying saucer' situation is not all imaginary or seeing too much in some natural phenomena. Something is really flying around." "The phenomenon reported is something real and not visionary or fictitious."[144]

In October 1947, a U.S. Air Force document classified "secret" included the statement: ". . . it is the considered opinion of some elements that the object *[sic]* may in fact represent an interplanetary craft of some kind."[145]

In the succeeding years, there were at least six U.S. Air Force projects and studies ostensibly aimed at solving the UFO mystery. Although these studies have been perceived by the news media and important segments of the public as having fully explained UFOs in "prosaic" terms, a closer study reveals their serious flaws and shortcomings. The following brief summaries describe the six studies.

Project Sign

Project Sign was the first U.S. Air Force investigation of UFOs and lasted from January 1948 to April 1949. Based at Wright Patterson AFB, Dayton, Ohio, it collected several hundred sighting reports from government and nongovernment sources, and claimed to explain most of them. Due to its unwillingness to accept UFO reports not sent *directly* to it, the Project Sign files include only a few dozen reports from 1947, while newspapers received more than 1,500 reports in just two weeks.

Project Grudge

Project Grudge replaced Project Sign in April 1949. In December 1949, a magazine article on UFOs written by the

---

[144] Air Force Base Intelligence Report, ibid.; Twining, ibid.

[145] Shulgen, Brig. Gen. George, ibid.

famous aviation writer Donald Keyhoe, based on his private investigations and military contacts, elicited enormous media attention. In it, Keyhoe insisted that UFOs were alien spacecraft and that the U.S. government was keeping this knowledge secret. In response to the furor that Keyhoe's article caused, and to demonstrate that there was nothing to get excited about, the Air Force reduced Project Grudge to a routine intelligence effort. However, in October 1951, Project Grudge was returned to its original status as a special project. This investigation ended in March 1952. The final report suggested that most sightings had been explained. However, a large percentage of the reports were left either unexplained or only conditionally explained.

Project Blue Book

The final open U.S. Air Force UFO investigation took over from Project Grudge in 1952 and lasted until December 1969. By this time, almost 13,000 sighting reports had been collected by all three projects combined. Approximately 600–700 cases remained unexplained (depending on which Air Force statistics are accepted). However, it is notable that hundreds of other cases have been labeled as explained without adequate justification and often in ways counter to known facts. Thousands of reports received conditional explanations (e.g., "possible balloon"; "probable aircraft"). But when the annual statistics were compiled, the qualifiers were dropped and "possible balloons" became definite balloons, as if speculative answers were established facts.

The project was closed down in late 1969, concluding that the continuation of Project Blue Book "cannot be justified, either on the ground of national security or in the interest of science . . . A panel of the National Academy of Sciences concurred in these views, and the Air Force has found no reason to question this conclusion." The memorandum recommending this action made it clear that the system which had long dealt with "reports of UFOs which *could* affect national security would *continue to be handled* through the standard Air Force procedures designed for this purpose," namely as it had all along—separately, "not part of the Blue Book system and in accordance with JANAP 146 or Air Force Manual 55-11."[146]

After the end of Project Blue Book, its case files were opened to public inspection at the Air Force Archives. They were withdrawn in 1974, to reappear in 1976 in the U.S. National Archives, after the names of all witnesses had been censored, thus preventing the reinvestigation of cases.

Project Stork

In late 1952, Project Blue Book director, Captain Edward J. Ruppelt, ordered a study of all the cases in the files for 1947–52, under a contract with the Battelle Memorial Institute. The data were supplied by the Air Force, while the conclusions were those of the Battelle scientists. The Air Force issued the final report as "Project Blue Book Special Report No. 14." It was released in 1955, accompanied by an Air Force news release. Although the Air Force stated their own conclusion that there was nothing to warrant interest or concern, this was contrary to the conclusions of the Battelle study. The Battelle scientists had stated that of almost 2,000 reports that were deemed to have sufficient information to permit analysis, 22.8 percent were judged to be "unexplained," and another 31.3 percent were judged to be "doubtfully" explained. In total, therefore, 54 percent of the sightings were said to lack convincing explanations![147]

---

[146] Bollender, Brig. Gen. C.H., ibid.

[147] Project Blue Book Special Report 14, ibid.

The Robertson Panel

In July 1952, a major sighting wave occurred and Air Force jet interceptors chased UFOs all over the country. Around Washington, D.C., UFOs were tracked by several radar installations simultaneously, and seen visually from the air and ground. The Central Intelligence Agency apparently became concerned that whether UFOs were real or not, the reports might clog the nation's intelligence channels, allowing an enemy to attack undetected.

In January 1953, the CIA convened a panel of scientists, chaired by H. P. Robertson, a scientist at Cal Tech, to look at some of the government's UFO data. The scientists were briefed by an Air Force team. After three days, the panel concluded that its original concern was correct, but that there was no convincing evidence that UFOs themselves were a threat to national security. The panel's recommendation that the government treat UFOs more openly was never implemented.[148]

The Condon Committee

By the mid-1960s the Air Force was becoming increasingly embarrassed by its poorly thought-out public statements on UFOs, which were highly criticized by the public. After Congressional hearings were held in response to public complaints, plans were begun to have one or more universities review the Air Force project and study the UFO situation independently. Eventually, the Air Force Office of Scientific Research gave a grant to the University of Colorado for a study to be headed by Dr. Edward U. Condon.

---

[148] Report of the Scientific Panel on Unidentified Flying Objects—Robertson Panel, January 17, 1953.

From the very beginning, the project was under a cloud of suspicion due to Dr. Condon's openly expressing his view in public forums that UFOs were nonsense.

A letter was discovered in the project files in which a prominent leader of the study suggested to university officials that skeptical scientists could be disarmed by assuring them that the study would only *appear* to be an objective one, and that the researchers were not expecting to find anything significant, in any case. Recently, further documentation has been found that makes it clear that the Air Force was encouraging the university to help them justify closing down Project Blue Book and abandoning open UFO studies.[149]

In early 1969, the $500,000 study was completed and the public received a strangely conflicting report, reminiscent of the conflicting Air Force/Battelle statements fifteen years earlier. Dr. Condon dismissed UFOs, reporting that they were without substance or significance. In the body of the report, however, more than 30 percent of the cases were left without satisfactory explanation. In some instances, the University of Colorado scientists made it clear that they were completely baffled by many things, such as sightings in which visual observations were confirmed by radar trackings.[150]

The Roswell Crash and Project Mogul

In July 1947, something crashed on a ranch outside of Roswell, New Mexico, giving rise to a long-term controversy: Was it an alien spacecraft, or a weather balloon as claimed by the U.S. government? At first, the U.S. Army Air

---

[149] Memo to E. James Archer and Thurston E. Manning from Robert J. Low, August 9, 1966; Letter to Dr. Condon from Lt. Col. Robert R. Hippler, USAF Science Division, January 16, 1967.

[150] Scientific Study of Unidentified Flying Objects, "Condon Report."

Force said it had recovered the remains of a "crashed flying disc," which was an early term for UFOs. This explanation was soon changed to "a weather balloon," which remained the official position for several decades.

Witnesses, interviewed long after the fact, describe debris of several types, including metallic materials of extreme strength and very light weight. They also tell of extraordinarily high security in connection with the recovery and shipment of a large quantity of debris (and in some versions, alien bodies) from the crash site.

In 1994, the U.S. Air Force announced that, in fact, the debris was from a cluster of balloons being tested for a long-secret project called Project Mogul, designed to detect nuclear explosions within the Soviet Union.[151]

In 1995, the Air Force published an exhaustive, thousand-page report documenting the Mogul balloon project, purporting to prove the Mogul explanation for Roswell. Once again, the Air Force summary conclusions conflicted with information in the body of the report. The data clearly indicate that the only test balloon clusters that conceivably could have landed at the ranch near Roswell were never tracked, so their landing sites remain unknown. Moreover, the balloons consisted only of familiar materials (not exotic metals) and would have quickly decomposed in the hot sun. The balloon clusters were held together by a braided line. The debris described by witnesses at the scene included neither braided line nor standard balloon material.

Conclusions

While the U.S. Air Force unquestionably had the capability to investigate UFOs scientifically, there is no evidence

---

[151] Weaver, Col. Richard L., USAF, ibid.

that it has ever done so. Published reports and related documents suggest studies that were hastily done, each time forced by short-term political considerations and public pressures rather than scientific inquiry. The resulting studies were superficial at best, inept at worst.

At various stages of UFO history, the Air Force high command considered UFOs to be possibly extraterrestrial spaceships . . . and at the other extreme an annoying public-relations problem. Even Air Force officers at the Air University, Air Command and Staff College, wrote reports puzzling over the Air Force position and raising serious issues about the significance of UFO data.[152]

---

[152] Air Force Research Reports on UFOs, Fund for UFO Research, 1995.

CONGRESSIONAL HEARINGS ON UFOs

Only two formal hearings on UFOs have ever been held:

I. The House Armed Services Committee convened the first hearing in 1966 in response to widely publicized sightings and strong public and editorial criticism of the handling of the Air Force Project Blue Book UFO program. This effort was supported by the House Minority Leader, Gerald Ford (R-Mich.), whose home state was the focus of many sightings.

Only witnesses connected to the Air Force project testified. Thereupon, the Secretary of the Air Force announced the formation of an outside review of Project Blue Book and an independent study of current cases. This resulted in the University of Colorado "Scientific Study of UFOs," which became known as the Condon Committee project, after the name of its director.

April 5, 1966. House Armed Services Committee (89th Congress, 2nd Session). Committee Print No. 55. "Unidentified Flying Objects."

Chairman: L. Mendel Rivers (D-S.C.)
Witnesses: Harold Brown, Secretary of the Air Force
Dr. J. Allen Hynek, Scientific Consultant to the Air Force
Maj. Hector Quintanilla, Jr., Chief, Project Blue Book

II. The House Science and Astronautics Committee convened a second hearing two years later, during the final stages of the Condon Committee project, to review the scientific evidence for UFOs. It took the form of a scientific symposium in which six scientists testified and six others submitted prepared papers.

July 29, 1968. House Science and Astronautics Committee (90th Congress, 2nd Session). Committee Print No. 7. "Symposium on Unidentified Flying Objects."

Chairman: George P. Miller (D-Calif.)
Hearing Chairman: J. Edward Roush (D-Ind.)
Witnesses: Dr. J. Allen Hynek, Head, Dept. of Astronomy, Northwestern University
              Dr. James E. McDonald, Senior Physicist, Institute of Atmospheric Physics, University of Arizona
              Dr. Carl Sagan, Dept. of Astronomy, Cornell University
              Dr. Robert L. Hall, Head, Dept. of Sociology, University of Illinois–Chicago
              Dr. James A. Harder, Assoc. Professor, Civil Engineering, University of California–Berkeley
              Dr. Robert M. L. Baker, Jr., Professor, Dept. of Engineering, University of California–Los Angeles

(Submitted statements from: Dr. Donald Menzel, Harvard College Observatory; Dr. R. Leo Sprinkle, Psychology, University of Wyoming; Dr. Garry C. Henderson, Senior Research Scientist, General Dynamics; Stanton T. Friedman, Nuclear Physicist, Westinghouse Astronuclear Laboratory; Dr. Roger N. Shepard, Psychology, Stanford University; and Dr. Frank B. Salisbury, Plant Sciences, Utah State University.)

Findings of the Condon Committee

In 1969, the Condon Committee published its findings. The project director, physicist Dr. Edward U. Condon, concluded that there was no convincing scientific evidence for UFOs and therefore recommended termination of Project Blue Book.

However, critics of the Condon Report pointed out that up to *30 percent of the cases investigated by the committee had remained unexplained*! According to the critics, such as Dr. J. Allen Hynek, Dr. Condon's conclusions were politically oriented rather than scientific (i.e., the Air Force wanted Blue Book terminated and needed a good reason).

Opinions of the Scientific Symposium

Of the six scientists who testified in the Symposium, five were of the opinion that there was a valid scientific anomaly that should be further investigated. Only Dr. Sagan, while conceding that some cases remained unexplained, was more skeptical. In fact, Dr. McDonald's thoroughly prepared paper with case histories is considered a milestone in UFO research. McDonald concluded: "My own study of the UFO problem has convinced me that we must rapidly escalate serious scientific attention to this extraordinarily intriguing puzzle."

Dr. Baker, whose testimony highlighted the unexplained nature of UFO movie films he had analyzed, recommended: "[establishment of] an interdisciplinary, mobile task force or team of highly qualified scientists . . . on a long-term basis, well funded, and equipped to swing into action and investigate reports on anomalistic phenomena. . . . We must get a positive scientific program off the ground. . . ."

Unfortunately, to date no such officially funded and open investigation has been undertaken.

Note

In 1976, Marcia Smith, a specialist in aerospace with the Congressional Research Service, prepared a comprehensive report on UFOs entitled "The UFO Enigma." It was revised and updated by George D. Havas in 1983 into a 143-page Report No. 83-205 SPR. It contains sections on types of encounters, witness credibility, pre-1947 accounts, history of Air Force UFO investigations, international perspectives, appendices with selected case summaries, etc. "The UFO Enigma" provides a well-researched and unbiased overview of the phenomenon.

INTERNATIONAL AGREEMENTS
AND RESOLUTIONS

Although this Briefing Document contains a small sample of
UFO cases, the global nature of the phenomenon is shown
by its geographical distribution. The cases studied include:
Germany (foo fighters), Scandinavia (ghost rockets), several
regions of the United States (Alaska, Washington, Washing-
ton, D.C., Texas, New Mexico, and the northern tier near the
Canadian border), England (Suffolk), Canada (Manitoba),
Brazil, Spain (Canary Islands), Iran, France, Belgium, and
Russia. UFO cases can be easily found for the rest of the world.

While the air forces (and in some cases other military, intelligence, space, and/or scientific agencies) in these countries have dealt with the UFO problem at one time or another,
there is little evidence of any long-standing open international cooperation effort. However, some examples of bilat-
eral, regional, and global approaches have been found.

I. 1975: Bilateral: USA–USSR

A curious clause about "unidentified objects" exists in
the *Agreement on Measures to Reduce the Risk of Nuclear
War between the United States of America and the Union of
Soviet Socialist Republics*. The Agreement was part of the
policy of detente during the Nixon and early Brezhnev ad-
ministrations. It was signed on September 30, 1971, by Secretary of State William Rogers and Foreign Minister Andrei
Gromyko.

The Agreement has nine articles on issues such as informing each other "against the accidental or unauthorized
use of nuclear weapons under its control," notification in advance of missile launches that go beyond the national territory

of each country, and other measures of cooperation in order
to avert "the risk of outbreak of nuclear war." Article 3 reads:

> *The Parties undertake to notify each other immedi-
> ately in the event of detection by missile warning sys-
> tems of **unidentified objects** [emphasis added], or in
> the event of signs of interference with these systems
> or with related communications facilities, if such oc-
> currences could create a risk of outbreak of nuclear
> war between the two countries.*[153]

The interpretation of Article 3 as including the possibility
of UFO incursions seems inescapable. It is indeed reassuring
in view of the cases where UFOs hovered over military facilities with nuclear weapons (SAC bases in the United
States, NATO bases in England, missile bases in Russia). On
the other hand, attorney Robert Bletchman has pointed out
that "unidentified objects" (UOs) include non-UFO situations as well (such as an accidental overflight by a civilian
aircraft or a terrorist attack), but in the final analysis, *UOs do
include UFOs.* What degree of cooperation about UOs/
UFOs existed between the United States and the USSR (and
currently with Russia) is hard to say, but Article 9 stated:
"This Agreement shall be of unlimited duration."

II. 1977–78: Global: United Nations

In the mid-1970s, the Prime Minister of the new member
state of Grenada, Sir Eric Gairy, began a lobbying initiative
to incorporate the UFO problem into the United Nations
agenda. Prime Minister Gairy and UN Ambassador Welling-

---

[153] United States Treaties and Other International Agreements, volume 22, part 2, 1971, "Union of Soviet Socialist Republics, Measures
to Reduce the Risk of Nuclear War Outbreak."

ton Friday raised the UFO issue at a meeting of the thirty-second General Assembly Special Political Committee on November 28, 1977. Grenada was proposing the "establishment of an agency or a department of the United Nations for undertaking, coordinating and disseminating the results of research into Unidentified Flying Objects (UFOs) and related phenomena."[154]

Grenada made further statements on November 30 and December 6, 1977. As a result of this effort, at the 101st plenary meeting on December 13, 1977, "the General Assembly adopted Decision 32/424," which acknowledged "the draft resolution submitted by Grenada" and further stated that:

3. The General Assembly requests the Secretary-General to transmit the text of the draft resolution, together with the above-mentioned statements, to Member States and to interested specialized agencies, so that they may communicate their views to the Secretary-General.[155]

Secretary-General Kurt Waldheim duly forwarded "Decision 32/424" to the Member States by a "note verbale" on March 13, 1978. However, only three governments responded (India, Luxembourg, and Seychelles) and only two specialized agencies (International Civil Aviation Organization and UNESCO) replied with a flat "no comments to offer."[156] Not deterred, Grenada launched a new offensive during the thirty-third General Assembly.

---

[154] United Nations Office of Public Information, "Special Political Committee Begins Debate on UFO Item," November 28, 1977.

[155] United Nations General Assembly, Thirty-third session, Agenda item 126, "Establishment of an Agency or a Department of the United Nations for Undertaking, Co-ordinating and Disseminating the Results of Research into Unidentified Flying Objects and Related Phenomena," Report of the Secretary-General, October 6, 1978.

[156] Ibid.

A group of recognized experts was brought to testify before a Hearing of the Special Political Committee on November 27, 1978. Besides Sir Eric Gairy and Wellington Friday, the Hearing included testimony by Drs. Allen Hynek and Jacques Vallee, and a firsthand witness account by Lt. Col. Lawrence Coyne of the U.S. Army (Reserve) on the famous 1973 UFO-helicopter near collision case in Ohio (see *Quotations*, section on Military/Intelligence). A letter of endorsement by astronaut Gordon Cooper, who was then Vice President of Research and Development at Walt Disney Enterprises, was also read into the record (see *Quotations*, section on Astronauts).

At the 87th plenary meeting of the General Assembly on December 19, 1978, Decision 33/426 was adopted with the same heading to the previous Decision 32-424 cited above, "Establishment of an agency or a department of the United Nations for undertaking, coordinating and disseminating the results of research into unidentified flying objects and related phenomena." The "consensus text" informed in its Point 1 that the General Assembly had "taken note" of the "draft resolutions submitted by Grenada" and that:

> *2. The General Assembly invites interested Member States to take appropriate steps to coordinate on a national level scientific research and investigation into extraterrestrial life, including unidentified flying objects, and to inform the Secretary-General of the observations, research and evaluation of such activities.*

> *3. The General Assembly requests the Secretary-General to transmit the statements of the delegation of Grenada and the relevant documentation to the Committee on the Peaceful Uses of Outer Space, so that it may consider them at its session in 1979.[157]*

---

[157] United Nations General Assembly, Thirty-third session, "Decisions adopted on the reports of the Special Political Committee."

Point 4 finally stated that the Outer Space Committee would permit Grenada "to present its views" in 1979 and the Committee's deliberation would be included in its report to the thirty-fourth General Assembly. The Grenada initiative was gradually opening the door to UFO cooperative international investigation, but unfortunately the effort came to an abrupt halt when the Gairy government was overthrown by a Marxist revolution led by Maurice Bishop. The new government launched a publicity campaign to discredit Gairy as a believer in voodoo and flying saucers. Decision 33/426 was never implemented, but its mere existence provides a useful framework for any future initiative on the matter.

III. 1990–93: Regional: European Parliament

As a result of all the activity registered during the UFO wave in Belgium, the European deputy, Mr. Di Rupo, who served as Minister of Education for Wallonia (the French-speaking region of Belgium where the wave occurred), proposed a motion in 1990 to set up a "European UFO Observation Center" under the aegis of the Committee on Energy, Research and Technology (CERT). The Di Rupo motion proposed that this Center "should collect together the isolated observations made by members of the public and by military and scientific institutions and organize programmes of scientific observation."[158]

The matter was eventually entrusted to another Euro-deputy, Professor Tullio Regge, an Italian member of the European Parliament with a Ph.D. in physics, who released a "Draft Report" on August 17, 1993. Professor Regge sought

---

[158] European Parliament, "Draft Report of the Committee on Energy, Research and Technology on the proposal to set up a European centre for sightings of unidentified flying objects (B3-1990/90)," Rapporteur: Mr. Tullio Regge, August 17, 1993.

the advice of Jean-Jacques Velasco, who heads SEPRA (Service for Assessment of Atmospheric Re-entry Phenomena) at the French National Center for Space Research (CNES) in Toulouse, as the only official European organization with experience in UFO investigations. The section titled "Motion for a Resolution" further stated:

> *The European Parliament . . . proposed that SEPRA be regarded as a responsible partner of the EC [European Community] so far as UFOs are concerned and that it be given a statute enabling it to carry out inquiries throughout the Community's territory. Any additional costs which might arise as a result of SEPRA's increased role must be covered by agreements between the French government and the other EC Member States or, where necessary and with the approval of the governments involved, directly between SEPRA and other EC research institutes or organizations.[159]*

The section titled "Explanatory Statement" in Regge's report consisted of a seven-page discussion of the UFO subject covering the following scientific, sociological, and political items:

> *1. Military secrets; 2. Alien civilizations; 3. Supertechnologies; 4. The role of the mass media; 5. Various explanations; 6. Link between show business and sightings; 7. Analogy with group religious experiences; 8. The recent spate of sightings in Belgium; 9. Unknown atmospheric phenomena; 10. Interviews with witnesses; 11. Air forces in the EC; 12. Conclusions.*

---

[159] Ibid.

The tone of the report was very cautious and did not endorse the extraterrestrial hypothesis. However, the report did recognize that a small percentage of UFO cases remain unexplained and warrant further scientific attention. The section on "Various explanations" concluded:

> *A second conclusion is that the few remaining inexplicable sightings (about 4%) must for the time being be regarded as UFOs (unidentified flying objects) in the literal sense of the term. The lack, perhaps temporary or accidental, of an explanation in no way allow us to regard a sighting as certain proof or even an indication that aliens exist, with technological capabilities vastly superior to our own. However, scientists still have a duty to continue researching into these events in order to arrive at a satisfactory explanation.[160]*

Regge's final conclusion was to propose that SEPRA expand its UFO activities to cover all the EC Member States:

> *It might be worthwhile, however, setting up a central office to compile and collate information concerning UFOs throughout the EC. Such an office could help, first and foremost, to stem the flood of uncontrolled rumors that confuse the public and become a point of reference when, as very frequently happens, sightings are reported . . . Lastly, the office could have an invaluable role to play in exploring the existence and nature of rare meteorological phenomena and could draw on the support of existing organizations. Given that SEPRA has acquired considerable experience in this field, the logical and economical solution would be to assign it a Community-wide role and Community*

---

[160] Ibid.

status, thereby enabling it to conduct investigations and disseminate information through the EC.[161]

Unfortunately, the European Parliaments did not have the necessary votes to implement and fund Professor Regge's recommendations and so the matter lies essentially dormant for the time being. As with the General Assembly Decision 33/426, however, the Regge motion for a European UFO Center linked to SEPRA remains as a potentially useful framework should the political will change in the future.

---

[161] Ibid.

EXAMPLE OF AIR FORCE POLICY WHEN QUESTIONED BY CONGRESS REGARDING RESULTS OF ITS UFO INVESTIGATIONS

In a letter to Senator Patty Murray (D-Washington), August 25, 1993, the Air Force stated:

> *The Air Force began investigating UFOs in 1948 under a program called Project Sign. Later, the program's name was changed to Project Grudge and, in 1953, it became known as Project Blue Book. On December 17, 1969, the Secretary of the Air Force announced the termination of Project Blue Book . . . As a result of these investigations, studies, and experience, the conclusions of Project Blue Book were:* **1) no UFO reported, investigated and evaluated by the Air Force has ever given any indication of threat to our national security . . .** [Emphasis added.][162]

Compare this with the statement of General Carroll Bolender, USAF, in 1969 when recommending the closing of Project Blue Book (unclassified, but sixteen attachments "could not be found"):

> *Moreover,* **reports of unidentified flying objects which could affect national security** *are made in accordance with JANAP 146 or Air Force Manual 55-11, and* **are not part of the Blue Book system** . . . *However, as already stated, reports of UFOs which could affect national security would continue to be*

---

[162] August 25, 1993, letter to Senator Patty Murray (D-Washington), from the Air Force.

handled through the standard Air Force procedures designed for this purpose. [Emphasis added.][163]

The Air Force statement to Senator Murray is the truth, but not the whole truth. Project Blue Book did not handle the important material that would affect national security. But since the public and Congress do not know this, the impression is given that the Air Force never discovered anything of importance among its many thousands of UFO reports. The Air Force chose to keep some UFO investigations classified and not to inform Senator Murray or other legislators either of their existence or of the results of their inquiries.

---

[163] General Carroll H. Bolender, memo of October 20, 1969.

THE ROSWELL CASE

In the middle of the first major American wave of UFO sightings in 1947, while the country was intrigued by reports of strange disc-shaped craft flying erratically overhead, an unusual crash was reported on a sheep ranch northwest of Roswell, in central New Mexico.

The first government officials on the scene were Roswell Army Air Field intelligence officer Major Jesse Marcel and counterintelligence corps Captain Sheridan Cavitt. According to Marcel, they surveyed a large area littered with unrecognizable debris. After careful examination, they brought two carloads back to Roswell.

Roswell AAF Public Information Officer, Lt. Walter Haut, distributed a press release describing the U.S. Army Air Forces' acquisition of "the remains of a flying disc." The story spread quickly across the country and around the world. A few hours after the news release, the commander of the 8th Air Force, Brig. Gen. Roger Ramey, announced to a small press conference that there had been a mistake: The debris consisted of nothing more than the remains of a common weather balloon.

Newspapers and radio stations carried the story of the exciting discovery, and then of its mundane explanation. The "crashed flying saucer" story vanished, not to be heard of again for more than thirty years. UFOs continued to fill the skies and the public imagination, but the thought that one of them might have crashed was barely considered.

In the late 1970s, private UFO researchers began to raise questions regarding the weather-balloon explanation. First-hand witnesses, such as Major Marcel, described in detail the large quantity of completely unfamiliar materials that covered a vast area of fifty acres. As more was learned, the chances of the Air Force having correctly identified the debris seemed increasingly remote.

Major Marcel and other technically competent witnesses described metallic foil lighter than household aluminum, yet impossible to crease, puncture, cut, or burn. They also found slender I-beams, similarly light and strong, which carried undecipherable symbols embossed on their sides.

The first book on the subject was published in 1980: *The Roswell Incident*, by Charles Berlitz and William Moore. It made a strong case for the "Roswell crash" having been a UFO rather than a balloon, due to the nature of the recovered materials, their wide dispersal, and the behavior of the security-conscious military.

Interest grew and more investigators went to work, locating and interviewing additional witnesses. By the late 1980s, it had become the most thoroughly investigated and best authenticated of all reported UFO crashes. There would soon be four books and scores of papers and television programs devoted to this single episode.

In the early 1990s, the refusal of the Air Force to comment publicly on the growing dispute led to a formal request for information from Rep. Steven Schiff (R-NM), in whose district the crash had occurred. His inability to get a satisfactory answer from the Pentagon led him then to ask the General Accounting Office to conduct a search for official documents related to the event.

The first official statement by the Air Force in a quarter century came in September 1994. This report explained that the debris found at Roswell was from a crash of then secret constant-altitude balloons designed to carry scientific equipment to detect Soviet nuclear explosions. Test flights of clusters of these balloons launched from Alamogordo, New Mexico, were part of a classified program called Project Mogul (which never became operational).

In July 1995, the GAO reported to Rep. Schiff that it had been unable to find documents explaining what really happened in the desert in 1947. It concluded that many documents from the Roswell Army Air Force Base had been improperly destroyed, and that "the debate over what crashed at Roswell continues."

In September 1995, the Air Force released a thousand-page report reinforcing its position that a Project Mogul balloon cluster was responsible for all the furor. It never quite said that a Mogul balloon rig had crashed on the sheep ranch, only that this was a possibility.

In fact, there is no evidence in *any* official report that such a balloon came anywhere near the sheep ranch, only that two such clusters were never found, and thus *might* have landed there. At the same time, the Air Force discounted the possibility that the debris could have been the result of the crash of a military airplane, the impact of a test rocket or missile, or any sort of nuclear accident.

With the GAO stating it had found no evidence for a Mogul balloon, and the Air Force eliminating most other possible explanations, the crash remains that of an unidentified flying object.

CHARACTERISTICS OF
IFOs AND UFOs

Common characteristics of **IFOs** (identified flying objects), which are *naturally occurring* by day, include clouds (especially lenticular), flocks of birds, and atmospheric phenomena such as "sun dogs" and "mock sun." Common characteristics of IFOs, which are naturally occurring by night, include stars, planets (especially Venus), meteors, and atmospheric phenomena such at St. Elmo's fire and ball lightning.

Common characteristics of **IFOs**, which are *man-made* and observed by day include unusual planes, balloons of different types, and helicopters. Man-made IFOs observed at night are usually airplane lights, satellites, spotlights, and advertising on strips trailing behind planes.

Common characteristics of **UFOs** (unidentified flying objects) can be described in terms of their appearance, their behavior, and their unusual effects.

1. UFOs generally appear in four main shapes: the disc (or saucer), the enormous triangle, the cylinder, and the sphere. They often exhibit brilliant luminosity, illuminating the terrain beneath, and shooting out beams of light. These beams are sometimes truncated. There is often a pulsation of a full spectrum of colors. Frequently, they are surrounded by vapor, sometimes appearing in cloudlike lenticular forms. Often the objects are domed, have portholes, and have a metallic-looking, shiny, lightly colored surface.

2. The behavior patterns of UFOs vary. Often there is a prolonged hovering, which is almost motionless, followed by an extreme acceleration that is frequently straight upward. Another familiar flight pattern is an erratic, nonlinear, nonsmooth, zigzagging,

darting motion. Sometimes the vehicles seem to flip end over end, or to stand upright in flight, or to fall like leaves. They are able to make very sudden right-angle turns at enormous velocities. They exhibit supersonic speed with no sonic boom. In fact, there is rarely any sound, sometimes just a high-frequency, low-volume humming sound.

3. UFOs have unusual effects upon their immediate surroundings. Animals may behave strangely, often panicking, cows' milk production ceases, etc. Electromagnetic effects cause electrical malfunctions to occur—car engines stop running, as well as car radios, headlights, etc.—as soon as the UFO is in close proximity. Upon departure of the UFO, all systems often start up again on their own. Furthermore, there may be signs of radiation effects on human, animal, and plant life in the immediate area as well as feelings of cold or heat.

4. UFOs leave after-effects on the ground. Impressions are often left forming a geometric design. Soil and vegetation in the affected area is dehydrated and will not absorb water. Affected vegetation will not seminate or regrow for a long time.

TERMINOLOGY OF UFOs

The frequently used term Close Encounters was coined by the late Dr. J. Alan Hynek as part of a terminology designed to categorize different types of UFO experiences. There are six main categories:

1. Nocturnal lights.
2. Daylight discs.
3. Radar-visual.
4. CE-I (Close Encounter of the First Kind) to denote a close observation.
5. CE-II (Close Encounter of the Second Kind) to denote cases where physical evidence is left by the UFO, i.e., ground traces, electromagnetic effects on motors, physiological effects.
6. CE-III (Close Encounter of the Third Kind) to denote cases where occupants are reported in addition to the object.

Researchers have recently expanded this terminology to include CE-IV for alleged abductions.

RESOURCES

Reading

1) **1964:**   *UFO Evidence.* Report by Richard Hall, of NICAP (National Investigations Committee on Aerial Phenomena—now defunct). 746 classic cases from the 1940s, 1950s, and 1960s reported by pilots, engineers, scientists, etc. Well documented. NICAP files are now archived by CUFOS.

2) **1968:**   *Congressional Hearings.* July 29, 1968. Symposium on Unidentified Flying Objects, Hearings before the Committee on Science and Astronautics, U.S. House of Representatives, Ninetieth Congress. Included testimony and prepared papers by Drs. Hynek, McDonald, Carl Sagan, and other scientists. The very detailed and well-documented McDonald Report summarized the best documented UFO cases up to that time.

3) **1969:**   *Condon Report.* Scientific Study of Unidentified Flying Objects, conducted by the University of Colorado under the direction of Dr. Edward U. Condon. Includes 117 well-documented cases, reviewed by aeronautic experts, physicists, and astronomers. Dr. Condon's well-publicized introduction attempted to disclaim the UFO phenomena, yet 30 percent of the cases admittedly could not be explained by any "natural" theory. (Published by New York Times/Bantam Books, 1969.)

4) **1984:**   ***Clear Intent.*** The first book, by Lawrence
                Fawcett and Barry Greenwood, to publish the
                declassified UFO files of the U.S. govern-
                ment. (Reprinted by Simon & Schuster with
                the title *The UFO Cover-Up*, 1992.)

5) **1987:**   ***Above Top Secret.*** By Timothy Good. World-
                wide cases, very detailed and well docu-
                mented. Originally published in England.
                (American edition published by Quill William
                Morrow, New York, 1989.)

6) **1988:**   ***Uninvited Guests.*** By Richard Hall, Aurora
                Press.

7) **1990–96:** ***The UFO Encyclopedia,*** three volumes. By
                Jerome Clark, Apogee Books.

Catalogs

UFO Sightings

UFOCAT. Computer catalog of raw UFO sighting reports from around the world, initiated in the 1970s by Dr.
David Saunders. It has over fifty thousand reports from five
continents and is currently maintained by CUFOS.

Ground Traces

Catalog of UFO landing cases where plants and soil were
affected, initiated by Ted Phillips in the 1970s. It has approximately four thousand reports from several countries.
The last printed catalog was published by CUFOS.

Pilot Cases

NASA scientist Dr. Richard Haines has kept a computerized catalog of UFO sightings by military, civil, test, and private pilots since the early 1980s. More than thirty-six hundred cases have been logged, including many in which electromagnetic effects were also detected by the aircraft instruments. Dr. Haines has published several papers on aspects of his pilot UFO catalog.

Vehicle Interference Catalog

Mark Rodeghier of CUFOS compiled in 1988 a catalog of 441 UFO reports in which car engines, batteries, or radios malfunctioned in close proximity to a UFO. A preliminary listing was published by CUFOS.

Medical Injury Catalog

Engineer John Schuessler of MUFON has compiled a catalog of close encounters in which the witness(es) suffered physiological effects and/or injuries. Four hundred cases were contained in the 1995 version of the catalog. A sample of the Medical Catalog was published by Schuessler in the 1995 MUFON Symposium Proceedings.

CUFOS, FUFOR, AND MUFON

Center for UFO Studies
2457 West Peterson Ave.
Chicago, IL 60659-4118

The Center for UFO Studies is a nonprofit, tax-exempt organization founded by Dr. J. Allen Hynek in 1973. Its mission is the scientific collection, evaluation, and dissemination of information about the UFO phenomenon. CUFOS comprises an international group of scientists, academics, investigators, and volunteers and is a clearinghouse for the two-way exchange of information where UFO experiences can be reported and researched. It maintains one of the world's largest repositories of UFO-related data.

Fund for UFO Research
P.O. Box 277
Mt. Rainier, MD 20712

This nonprofit corporation was established in 1979 to raise money to support scientific and educational projects submitted by qualified researchers. It is composed solely of a fifteen-member Board, most of the members being Ph.D.'s in various scientific fields. In its sixteen years, it has raised more than $500,000, which has been used to fund investigations in the physical and social sciences, to support scientific conferences, and to encourage the serious treatment of UFOs by the press.

Mutual UFO Network
P.O. Box 369
Morrison, CO 80465

The Mutual UFO Network, Inc., is the world's largest UFO investigative and research organization, with representatives in thirty-nine countries and every state in the United States. It is a nonprofit, tax-exempt corporation dedicated to resolving the phenomenon of Unidentified Flying Objects through scientific investigations and research. The results of major cases are published in the *MUFON UFO Journal*, a monthly magazine. MUFON sponsors an annual international UFO symposium and publishes papers from the symposium proceedings.

SOURCE MATERIALS
(AVAILABLE AT THE FUND FOR UFO RESEARCH)

Fund for UFO Research
P.O. Box 277
Mt. Rainier, MD 20712
Tel: (703) 684-6032

DON BERLINER is an independent journalist and aviation historian who has been writing full-time for thirty years, after having been a newspaper reporter/photographer, assistant editor, and Capitol Hill correspondent for a group of scientific newsletters.

His specialized experience in the UFO field includes three years (1965–68) as a staff writer for the National Investigations Committee on Aerial Phenomena (NICAP), then the world's leading private UFO agency, and eighteen years on the executive committee of the nonprofit Fund for UFO Research, which he helped found in 1979.

Berliner's many books include *Airplane Racing, Helicopters, Record-Breaking Airplanes, Airplanes of the Future, Victory Over the Wind,* and the definitive report on the 1947 Roswell, New Mexico, UFO incident, *Crash at Corona* (with Stanton Friedman).

WHITLEY STRIEBER is the author of twenty-two books, including the bestselling books on the UFO subjects. Among these are *Communion, Transformation, Breakthrough, The Secret School,* and *Confirmation. Confirmation* was made into a two-hour special by NBC. Strieber is the host of the Sunday night radio program "Dreamland," which airs on 300 stations and has seven million listeners. He is the co-founder, with his wife, Anne, of the Communion Foundation, devoted to the study of the UFO enigma.